Earth Architecture

UNIVERSITY PRESS OF FLORIDA

Florida A&M University, Tallahassee
Florida Atlantic University, Boca Raton
Florida Gulf Coast University, Ft. Myers
Florida International University, Miami
Florida State University, Tallahassee
New College of Florida, Sarasota
University of Central Florida, Orlando
University of Florida, Gainesville
University of North Florida, Jacksonville
University of South Florida, Tampa
University of West Florida, Pensacola

University Press of Florida

Gainesville · Tallahassee · Tampa · Boca Raton · Pensacola · Orlando · Miami · Jacksonville · Ft. Myers · Sarasota

WITHDRAWN

William N. Morgan

Earth Architecture

FROM ANCIENT TO MODERN

Copyright 2008 by William N. Morgan
This book was published with the generous support of Furthermore:
a program of the J. M. Kaplan Fund.
Printed in China on acid-free paper

13   12   11   10   09   08   6   5   4   3   2   1

Library of Congress Cataloging-in-Publication Data
Morgan, William N.
Earth architecture : from ancient to modern / William N. Morgan.
p. cm.
Includes bibliographical references and index.
ISBN 978-0-8130-3207-8 (alk. paper)
1. Earthwork—History. 2. Earth construction—History. 3. Vernacular
architecture—History. 4. Fortification—History. I. Title.
TA715.M55 2008
721.'04492—dc22   2007038094

The University Press of Florida is the scholarly publishing agency
for the State University System of Florida, comprising Florida A&M
University, Florida Atlantic University, Florida Gulf Coast University,
Florida International University, Florida State University, New College
of Florida, University of Central Florida, University of Florida,
University of North Florida, University of South Florida, and
University of West Florida.

University Press of Florida
15 Northwest 15th Street
Gainesville, FL 32611-2079
http://www.upf.com

To Eduard F. Sekler,
Professor of Architecture, Emeritus
Harvard University, Graduate School of Design

Contents

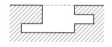
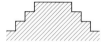
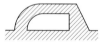

 **Water Retained** 133

 **Cities** 151

Foreword

Has Bill Morgan written a book on architecture or on landscape architecture? I believe the resonance and importance of this volume is the documentation of human-built examples that blends the two. With my own background in landscape architecture, I find *Earth Architecture* a fascinating documentary of how humans have reshaped the surface of the earth into places—earthworks that reflect the contemporary values, technologies, and functional needs of the cultures that build them. Architects, cultural geographers, archaeologists, and anyone interested in human ecology will be equally intrigued and inspired, for this work explains that these earthworks are all perspectives of the same story. It is a human story and, hence, a complex one. Our species has not been shy about modifying our environment to suit, with motivations that range from ceremony and defense to transport efficiency, resource extraction, and aesthetics, as well as for inexplicable reasons shrouded in the mystique of antiquity. Many of Morgan's examples inspire awe—because they were designed to do just that: massive burial memorials to revered and powerful leaders or sacred enclosures for worship that evoke the grandeur of the gods. Others amaze us with the engineering feats and human energy necessary to build them, leaving us to conjecture what ancient techniques were used and imagine the many life stories that were part of their realization and use. Still others are testament to the human brazenness with which at times we have used our planet, leaving great, gaping scars that could be seen from space long before we had the ability to do so.

Bill Morgan has compiled a categorical overview through the eyes of an architect of how humans shape earth into places— whether you call it architecture, landscape, agriculture, earthwork, or art. His carefully chosen aerial images, reconstructed plans, and diagrams are the architect's visual language that simplify the story and help us see the universality of earth building. They are geographically organized thought—categorized and thoroughly explored—that clears away the thickets and rubble covering the ruins and fills the gaps eroded by time, but without the grids and dry data of archaeological reports. The lessons are clearly distilled and easily reinterpreted to inspire modern forms that are expressions of our contemporary cultures.

Indeed, though this intriguing work may be the culmination of Morgan's fifty years of the subject's study, the lessons are more relevant for today and for the future than ever. Sustainability has become a driving value, even an imperative, for those of us who plan and design the built environment. There are reminders here of what it looks like when we build without concern for ecological or social impact, what we need to avoid, and what we need to do better. More inspiring, there are ancient and current examples here of how to use earth as a material that contributes to sustainable places by utilizing the earth's insulation and thermal mass to temper heat and cold, by taking advantage of this non-toxic natural material with low-embodied energy, and even by integrating habitat or storm-water management. The use of the site's earth inherently captures a genius loci, literally embedding, or at times extruding, architecture from place. Earth architecture blends architecture and landscape, functionally and symbolically integrating our dwellings with our planet, reminding us to think holistically with building design.

Joe Brown
President and CEO of EDAW

Acknowledgments

My research began some fifty years ago while I was a student attending the Harvard University Graduate School of Design. Over the years since then, my study has been encouraged and guided by Eduard F. Sekler, now Professor of Architecture Emeritus of Harvard University, to whom I owe my very special gratitude. At the outset of my research, the late professor Josef Zalewski also was most helpful by drawing my attention to such remarkable earthworks as the earth memorial to Tadeusz Kosciuszko near Kraków, and by urging me to work with Dr. Stephen J. Poulos of the Harvard Soil Mechanics Laboratory with the view of understanding more thoroughly the potentials of the earth (see appendix).

The Arthur W. Wheelwright Traveling Fellowship of the Harvard University Graduate School of Design made possible my travels to many sites in both the Eastern and Western hemispheres. During my visit to Cambodia, I serendipitously received a remarkable site plan and aerial perspective of Preah Vihear from architect J. Dumarcay of L'Ecole Française d'Extrême Orient, for which I am most grateful.

During the 1960s and 1970s, my research and data acquisition for the study, now called *Earth Architecture,* increased steadily. Of great assistance was a generous grant form the Graham Foundation for Advanced Studies in the Fine Arts under the very capable guidance of Carter Manny Jr. Arthur Drexler, then director of Architecture and Design at the Museum of Modern Art in New York, became interested in the project and introduced me to Ludwig Glaeser, curator of the Mies Van der Rohe Archives at MOMA, whose access to design resources was exceptionally helpful. Assisted by researcher Penelope Ray, our data bank expanded to between four hundred and five hundred examples of earth architecture worldwide. Critical to our research effort were the Avery Museum of Columbia University, the New York City Library, and the Smithsonian Institution in Washington, D.C.

As our files increased, our funds decreased, difficulties in organizing the immense quantity of materials multiplied, and enthusiasm for the project periodically lagged. As time permitted, I began to concentrate my efforts on the limited part of the assembled data that related to pre-Columbian architecture in the eastern United States. Impressed by my reconstructed site plans of ancient sites at comparable architectural scales, Philip Johnson guided me to an appropriate publisher. Steven Williams and Jeffrey

Brain of the Peabody Museum at Harvard University and J. C. Dickinson Jr. and Jerald Milanich of the Florida Museum of Natural History assisted me by patiently elevating my scholarship to an acceptable level. Watson Break, Adena, Newark, Poverty Point, Cahokia, and Emerald Mound are examples of sites in *Earth Architecture* that are based on this phase of my research effort, assisted by Joe Saunders, Jon L. Gibson, Melvin L. Fowler, and many other knowledgeable archaeologists.

Several years later I assembled and published a second study of pre-Columbian architecture in North America, this time in the geographical area of northwestern Mexico and the southwestern United States. Paquimé, Casa Grande, Taos, and Shipaulovi are sites from my earlier publication that reappear in a new format in *Earth Architecture.* Of great importance to this phase of my research have been Roger Kennedy of the National Museum of American History, W. Richard West Jr. of the National Museum of the American Indian, Bruce D. Smith and Karen Dohn of the National Museum of Natural History, Joseph J. Snyder, John B. Carlson, Stephen Lekson, Rina H. Swentzell, Cathy Cameron, John Stein, Jack E. Smith, Peter J. McKenna, Keith M. Anderson, Arthur H. Rohn, Todd W. Bostwick, Jeffery S. Dean, Steve Germick, George J. Gumerman, Frank W. Eddy, Robert C. Savey, Douglas W. Schwartz, Mark Michel, David R. Wilcox, Alphonso Ortiz, Stephen Athens, and especially George Anselevicius, then dean of the University of New Mexico School of Architecture and Planning.

A USA Fellowship from the National Endowment for the Arts made possible extensive travel and site visitations in the Southwest. I am grateful to Randolph McAusland and Mina Berryman of the NEA Design Arts Program for their patient assistance.

Special thanks are due to Thomas F. Harkins of the Duke University Library for photographs of the Duke Stadium; the Centre de Recherche Français de Jérusalem for material regarding Safadi; to the International Society for Educational Information Press of Tokyo and architect Ryosuke Yoshioka for illustrations of Emperor Nintoku's tomb; to Hamish Macdonald of the University of Auckland Anthropology Department and archaeologist Nigel Prickett for information regarding the Maori earthworks of New Zealand; the Bernice P. Bishop Museum of Honolulu for information and material relevant to the Marae of Tahiti; to D. K. Mukai and Ed Matsura of the R. M. Towill Corporation of Honolulu for the photogrammetric survey of Babeldaob; to Douglas Osborne, George Gumerman, et al. for information regarding Palauan terraces; and to the following architects, landscape architects, planners, engineers, sculptors and others whose creative works appear in *Earth Architecture*:

Atelier 5, architects: Siedlung Halen, Bern, Switzerland

Van Der Broek en Bakema, architect: Pampus Plan, Amsterdam, Netherlands

Harlan Bartholomew and Associates, planners: Bay of Naples, Florida

Bechtel Corporation, engineers: Nacimiento Dam, California

Max D. Lovett, architect/CRS: San Angelo Stadium, Texas

Le Corbusier, architect: Gardens of the Governors' Palace, Chandigarh, India

Ernst Cramer, landscape architect: Poet's Garden, Zurich, Switzerland

Hassan Fathy, architect: New Gourna Village, Egypt

Wallace Holm and Associates, architects: Patton School, Monterey, California

Frank Gehry, architect: Concord Pavilion Amphitheater, Concord, California

Jorge Romero Gutiérrez, architect: Helicode de la Roca Tarpeya, Caracas

Philip Johnson, architect/WDD: Hendrix College Library, Conway, Arkansas

Rick Joy, architect: Studio and Courtyard, Tucson, Arizona

Nader Khalili, architect: Eco-Dome, Hesperia, California

Ricardo Legoretto, architect: Beach Cabañas, Baja California Sur, Mexico

Maya Lin, sculptor: Vietnam Veterans Memorial, Washington, D.C.

Meyers and Bennett Architects/BRW: East Bank Bookstore and Offices, University of Minnesota

Gyo Obata, architect: Dallas–Fort Worth Airport, Texas

Caesar Pelli and A. J. Lumsden, architects/DMJM: Urban Nucleus, Santa Monica, California

Thomas Phifer, architect: Spencertown House, Hudson Valley, New York

Arnaldo Pomodoro, sculptor: New Cemetery proposal, Urbino, Italy

John Ray, architect: Dam Town proposal, near Pikeville, Kentucky

Richardson, Severens & Scheeler, architects: Undergraduate Library, University of Illinois, Urbana

Senmut, architect: Queen Hatshetsup's Mortuary Temple, Deir el Bahri, Egypt

Paolo Soleri, sculptor: Visions of Paolo Soleri, Scottsdale, Arizona

Hugh Stubbins, architect: Pusey Library, Harvard College, Cambridge, Massachusetts

Horace Trumbaner, architect: Duke University Stadium, Durham, North Carolina

Frank Lloyd Wright, architect: Second Jacobs House, Middleton, Wisconsin

A special word of appreciation is due to my son, W. Newton Morgan Jr., for his splendid photography of ancient sites, particularly in North America and Micronesia; without these images this study would remain unfinished.

Finally I must acknowledge that this book would not have been possible without the patient and tireless efforts of my wife, Bunny, who has managed to keep the project organized and moving forward through decades of seemingly endless revisions and corrections, more often than not with a sense of humor and always with a belief in the underlying value of the study.

While assistance from many sources must be acknowledged, the author alone assumes full responsibility for the final outcome of the study.

Earth Architecture

Middleton
Minneapolis
Cahokia
Miamisburg
Newark
Poverty Point
Scottsdale
Taos
Hudson Valley
Concord
Washington, D.C.
Nacimiento Dam
Williamsburg
Urban Nucleus
Durham
Hesperia
Dam Town
Expressways
Charleston
Coolidge
Natchez
Tucson
Mound Key
Cabo San Lucas
Dallas - Fort Worth
Paquimé
La Venta
Uxmal

Glaumbaer

Tahiti

Machu Picchu

New Plymouth
Auckland

Western Hemisphere Earthworks

Polders
Trelleborg
Waterloo
Avebury
Urbino
Guadix
Tivoli
Epidaurus
Matmata
Deir el Bahri
Djenné
Esfahān
Persepolis
Safadi
Katsura
Zhog Tou
Sakai
Shikoku
Preah Vihear
Ifugao
Corregidor
Angkor
Babeldaob
Borobudur

Eastern Hemisphere Earthworks

Introduction

Earth turns to gold . . .
In the hands of the wise.

RUMI (1207–1273 A.D.)

Earth Architecture is a study devoted to the architectural uses of earth in shaping the environment of humankind, a subject closely related to human ecology. Some fifty years in preparation, the study contains numerous precedents for sustainable design, energy conservation, and environmental adaptation. *Earth Architecture* includes contemporary as well as historical and vernacular examples drawn from many cultures and periods. The survey ranges in scope from subterranean dwellings to large-scale engineering projects with the view of exploring the formal and practical potential of earth shaping.

Structures built of earth presently house an estimated 1.5 billion people, about 30 percent of the world's population (Keefe 2005). Archaeologists have found evidence of mud-brick buildings constructed as early as ten thousand years ago in the Middle East and North Africa, where impressive buildings up to ten stories high have been recorded in an unbroken architectural tradition that continues today. Although creating individual earth structures is a familiar practice in many areas of the world today, the practice of reshaping the earth to create new human environments is little known.

My interest in the relation of architecture to environmental adaptation began in the 1950s. An architectural problem when I was a student at the Harvard Graduate School of Design called for a museum to commemorate the exploration and settling of the American West. The site consisted of a broad, unspoiled plain bounded by a distinctive line of hills in the limitless expanse of the Southwest. From a design point of view, placing a conventional museum building *on the site*, no matter where, seemed inappropriate both to the site and to the spirit of the memorial. After weeks of failed attempts to design a building on the site, I realized that the appropriate memorial was *the site itself*. The project then developed with grace and ease: a formal plateau well up the steep hillside became the site of a truncated pyramid containing fragile exhibits in a glazed court. The plateau's upslope wall was ornately sculpted with bas-reliefs and inscriptions, reinforcing the spirit of the memorial and enhancing its special sense of place.

Years later when I visited Monte Alban in Oaxaca, I had an unusual sense of having been there before. Most of the materials that I proposed for the construction of the projected hillside memorial were recycled earth materials taken from the site and returned to it in new configurations. Although most of the faculty reviewers had reservations about accepting a student proposal to be constructed of dirt, one professor, Joseph Zalewski, encouraged my further exploration. He drew my attention to the man-made mountain of earth created near Kraków in the early 1820s as a monument to the Polish national hero Tadeusz Kosciuszko, who also was a hero in the American Revolution. After Kosciuszko's death in 1817, people from every part of impoverished, war-ravaged Poland brought earth to his resting place to establish a lasting memorial in their land. The enduring earth memorial survives today in very nearly its original form.

In 1964 an Arthur W. Wheelwright Fellowship afforded me the opportunity to travel to a number of sites of earth architecture around the world, to talk with knowledgeable scholars about the earth, and to expand my understanding of the subject in new directions. Fellowship funds made possible tours of the Western Hemisphere in 1964–65 and of the Eastern Hemisphere a year later. Visits to many of the sites were planned, but other sites were found entirely by chance.

In Peru, while I was awaiting transportation to Cuzco and Machu Picchu, an enterprising cab driver learned of my interest in earthworks. He drove me to the remains of Pachacamac, an Inca monument on the banks of the Lurín River some eighteen miles south of Lima on the Pacific Ocean. The scale is vast: the coastal mountains form an abrupt wall on the east side of the arid plain facing the seemingly endless expanse of the Pacific Ocean. In the center of the site a mountain juts up along the otherwise featureless coastline on the edge of the sea. On closer inspection I realized that the mountain, perhaps originally a natural feature in the landscape, is in fact an immense complex of terraces, ramps, walls, portals, and other features said to have been at one time an Inca Temple of the Sun. Even much in ruins, Pachacamac is one of the most compelling architectural monuments in the New World.

Continuing my Western Hemisphere tour, I visited the Maya ceremonial centers at Chichen Itzá and Uxmal, and proceeded to Villahermosa to arrange for a flight to the mountainside site of Palenque. At an archaeological park in Villahermosa I discovered an elusive reference to La Venta, a very early site that seemed to lie at the dawn of pyramid building in Mesoamerica. After searching for definitive information about La Venta for several years in both Mexico and the United States, I eventually obtained enough information to propose the reconstructed drawing of the Olmec pyramid and plaza shown in the "Mounds" section of this study. The fragile pyramid of earth clearly demonstrates an exuberant spirit of architecture for the succeeding generations of builders who learned to protect their structures with stone cladding. La Venta also features an extensive site axis, a clearly defined plaza, and related architectural features that anticipate later, more highly developed Mesoamerican pyramid complexes of stone.

While seeking information about La Venta, I happened upon references to recent excavations revealing the earth architecture of Cuicuilco, an ancient earthwork underlying present-day Mexico City. Cuicuilco was apparently a major element in the early pre-Columbian town that predated Tenochtitlán. A photograph of the remains appears in the "Cities" section of our study.

After brief visits to the adobe pueblo of Acoma in central New Mexico and to central Arizona for an interview with Paolo Soleri, I began the Eastern Hemisphere tour of my Wheelwright Fellowship early in 1966. At an open market in central Phnom Penh, a merchant gave me change for a minor purchase in the form of a Cambodian banknote. Engraved on the banknote was the representation of an extraordinary earthwork, an extended ceremonial way traversed by elephants that ascended a gently sloping hillside to a dramatic cliff overlooking a vast expanse of jungle lowlands. My interpreter advised me to inquire about the site at L'Ecole Française d'Extrême Orient in Siem Reap, where Angkor Wat and the Baphuon are located. Here architect J. Dumarcay not only explained the name and structural system of Preah Vihear but also presented me with a very accurate copy of the site plan and perspective that I had seen on the banknote. The French scholars were in the process of hurriedly evacuating their archaeological projects at Angkor in anticipation of acts of barbarism, which unfortunately did occur shortly after we left Siem Reap.

What I did not know about Preah Vihear is that the site was at that time the focus of a border dispute between Cambodia and Thailand. As I stepped off the airplane in Bangkok, I carried my accurate site and topographic plan of Preah Vihear in a roll under my arm. The customs guards immediately seized my map and threw my wife and me into a barren, dark room. After several hours and repeated protests, a U.S. embassy representative arrived and arranged for us to board the next plane for India with all our possessions except the beautiful engravings of Preah Vihear.

One of the most remarkable earthworks that I discovered in gathering materials for this study is the Gandhi Memorial in New Delhi. The site lies on the silted plain of the Yamuna River about half a mile southeast of the Red Fort. The memorial essentially consists of a truncated pyramid formed by four very gently sloping, grassy sides. In keeping with the spirit of the memorial, the earth construction materials for the pyramid were placed entirely by hand, and the maintenance of the grounds is performed entirely by hand as well. Four walks enter the geometrically precise truncate from the cardinal points. The walks proceed between stone-lined walls of increasing heights until they enter the truncate below a circumambulatory crowning the earthwork. Upright stone slabs block the views of the interior courtyard as visitors approach. A black marble slab in the courtyard's center marks the place of the cremation of Mahatma Gandhi's remains; his ashes were scattered in the Yamuna River. When I came upon the memorial early in 1966, it was nearing completion. Also nearing completion nearby as part of the memorial was a children's play park with shade trees, pools, and benches.

Having visited sites in Fatehpur Sikri, Agra, New Delhi, and Chandigarh, I proceeded to Isfahan to visit the beautiful Khajou Bridge, a multipurpose structure that is discussed in greater detail in the "Water Retained" section of this study. At this point I must explain to the reader that, as part of my study, I collected soil samples with the view of being able to determine with scientific accuracy certain facts about specific sites on a comparative basis. The soil samples were collected from undisturbed material about three feet (one meter) below the surface vegetation. The uniform samples were stored in small sealed tubes for later analysis. Since the Khajou Bridge site is subject to periodic flooding, I decided to take a sample along the shore on the downstream side of the bridge approach, where silting was not a factor. My intentions were well meaning, and all in the interest of science, of course. Some months later the preliminary analysis of my soil sample for the Khajou Bridges indicated that I had obtained an unadulterated sample of roughly two-hundred-year-old donkey droppings, most likely swept over the side of the bridge then as now. So much for my scientific research.

After visiting Persepolis, Petra, Deir el Bahri, and several other sites, I returned to the United States to organize my data. During an interview in Philadelphia, Louis Kahn shared his thoughts with me about the creative potentials of the earth and told me of an extraordinary project he imagined while working on his designs for Dacca: he envisioned a hillside community consisting of trapezoidal compounds defined by earth-berm walls. In plan, the base of the community was a riverside esplanade from which paved streets ascended the hillside diagonally so that each compound had streets on all four sides. The streets sloped continuously and uniformly with gradients that were not too steep for pedestrians. Utility lines were recessed into the earth berms but open to the sky; they distributed fresh water from the hilltop by gravity and conveyed away effluents down the slope. Extending the diagonal street pattern along the riverbank and following the contours of the terrain along the hillside without compromising the plan's geometry would accommodate future expansion. We concluded our day-long interview with lunch around four o'clock by strolling down Walnut Street for a small box of tomatoes at the Farmers' Market.

Kahn's vision for the hillside community near Dacca demonstrated not only his appreciation of the earth's potential to shape the human environment but also of East Pakistan's urgent need to employ large numbers of its unemployed and largely unskilled workers, to house them decently, and to live up to the citizens' aspirations for their new nation.

By 1967 I had developed nine basic categories for the organization of a comprehensive study of earth architecture, together with distinctive icons representing the categories. *Progressive Architecture* published my organizational table, listing some fifty-eight examples of earth architecture in its April 1967 issue, and devoted its monthly publication to matters relating to earth architecture: earthmoving equipment; excavation techniques including blasting; historical sites of caves and troglodytes; military and naval installations; commercial and institutional facilities utilizing the earth; soil mechanics; dams, reservoirs, and canals; and related earthwork, together with commentaries and discussions by various authorities. The publication validated much of my study and opened new areas for exploration.

Exhibitions and publications by the Museum of Modern Art were a source of encouragement for my ongoing studies of the earth in the 1960s and 1970s. For several years I worked in New York with Ludwig Glaeser on my research for the earth project. A generous grant from the Graham Foundation for Advanced Studies in the Fine Arts brought fresh impetus to our research effort. Together with our research assistant, Penelope Ray, we gathered examples of earth architecture from the New York City Library, MOMA research files, the Library of Congress, the Harvard Graduate School of Design Library, the Smithsonian Institution Anthropological Archives, and many other sources. In the course of many months, our research effort yielded some fifteen file boxes filled with uncorrelated data illustrating many of the sites that the reader will find in this study.

As funding for our research effort dwindled, the demands of my fledgling architectural practice intensified. The design of a museum of natural history on a hillside site offered me the opportunity to create terraces and berms on a scale in which I had not worked previously. Interviewing the scholars who would utilize the new museum, I met Ripley P. Bullen, a revered archaeologist on whose cluttered desk lay a drawing of an earthen pyramid with outlying architectural features. To my surprise, he explained that the site was situated not in Mesoamerica but in Florida. Ancient architecture in Florida? I was amazed. For three days I searched through the clutter in disbelief. Shortly thereafter I met archaeologist Jerald T. Milanich, a particularly knowledgeable scholar on matters pertaining to Florida and the Caribbean.

For a comprehensive overview of pre-Columbian North America, I was directed to Stephen Williams of the Peabody Museum of Harvard University, who generously guided my research and introduced me to numerous authorities in the field over a period of several years. My efforts eventually led to the publication of *Prehistoric Architecture in the Eastern United States* (MIT Press, 1980) and *Precolumbian Architecture in Eastern North America* (University Press of Florida, 1999). My interest in the origins of architecture and my earlier experiences as a naval officer in the Mariana Islands converged in *Prehistoric Architecture in Micronesia* (University of Texas Press, 1988). Visits to several ancient sites in New Mexico and the prompting of my associates led to *Ancient Architecture of the Southwest* (University of Texas Press, 1994). These four books owe much to my earlier research in earth architecture.

Returning to *Earth Architecture* in 1994, I sought the advice and guidance of Eduard Sekler, my professor of architecture at the HGSD who had introduced me to architectural history in the 1950s and has encouraged me throughout my ensuing years of practice. After reviewing an interim summary, Eduard recommended eliminating such rock-cut sites as Haibak, Ajanta, and Ellora in favor of more yielding earth materials. He also suggested several other improvements, such as the inclusion of examples of rammed earth and adobe, materials utilized skillfully by Egyptian architect Hassan Fathy. Perhaps the most difficult task in organizing the work was reducing the hundreds of well-illustrated examples in our data bank to the ninety earthworks presented in this study.

During the fifty years or so since I began this study, the way that I think about architecture and the environment seems to have shifted. During the 1950s buildings appeared to exist independently from their sites; structures could be reoriented or relocated largely at will. Frank Lloyd Wright's works respected nature without reservation, but seemingly at a distance; and Le Corbusier elevated his structures on stilts so that they safely maintained their detachment from nature. The conception of architecture as a relocatable object in the landscape may have its origin at least in part in our classical Greco-Roman tradition. Gradually my interests have shifted from detachable architecture to structures more closely integrated with their environments. Visits to ancient sites in Egypt and the Middle East as well as research into the origins of architecture in North America turned my attention increasingly toward an architecture in which earth is a major building material and platforms, mounds, shafts, and terraces are integral architectural elements of the environment. *Earth Architecture* suggests an ecologically sustainable relationship between human beings and their environments, a more enduring relationship rooted in place and in meaning through time.

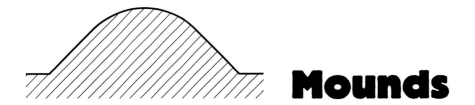

Mounds

Avebury Circle, 2600–1800 B.C., Wiltshire, England

Olmec Pyramid and Plaza, La Venta, 800–700 B.C., Tabasco, Mexico

Adena Burial Mound, circa 100 B.C., Miamisburg, Ohio

Hopewell Octagon and Circle, circa 200, Newark, Ohio

Trelleborg, Viking Fort, circa 1000, Slagelse, Denmark

Lion Mound Memorial, 1815–1826, Waterloo, Belgium

M OUNDING THE EARTH to create artificial hills or bermed enclosures seems to have been one of the earliest methods of making architectural statements that subsume the ideas, concerns, and ambitions of their builders. For example, between 3900 and 3300 B.C. Native Americans began to construct conical earthworks up to 24 feet (7.5 meters) high along the shores of the Ouachita River in Louisiana. Now known as Watson Brake, the remarkable monuments apparently served primarily to identify territories in the wilderness. In time the builders interconnected the distinctive mounds with low terraces forming an unprecedented oval enclosure measuring at most 800 by 1,000 feet (240 by 300 meters).

In a different setting altogether, construction began around 2600 B.C. in England on one of the foremost monuments of ancient Europe. In time an immense grass-covered embankment almost 1,400 feet (428 meters) in diameter was completed with an inner ditch and entrances at the cardinal points. Now encircling the village of Avebury, the original purpose of the great earth enclosure is the subject of continuing speculation.

Between 3900 and 3300 B.C., earth mounds were shaped to form an oval enclosure known as Watson Break in the wilderness of north central Louisiana. Drawing by Jon Gibson.

In honor of a national hero, citizens brought earth from distant parts of Poland to build the Kosciuszko Memorial near Kraków between 1820 and 1823. Photo by H. Hermanowicz.

Sometime after 800 B.C. in Mesoamerica, Olmec builders erected an impressive clay pyramid and plazas at La Venta, a structure anticipating the later stone-encased pyramidal hills of the Maya, Toltec, Zapotec, and neighboring groups. Although pyramids and plazas in pre-Columbian Mexico seem to reflect the prestige and aspirations of the builders, conical hills in the pre-Columbian Ohio Valley more often were related to burial ceremonialism. At such Native American sites as Miamisburg, Adena people constructed often immense earth mounds above the log tombs marking the final resting places of revered leaders.

Toward the beginning of the Christian era, immense geometric enclosures appeared at Newark and other Native American sites in the Ohio Valley; here ceremonialism was a central concern of the Hopewell people, who often created vast earthworks where large numbers of participants could assemble. While great geometric enclosures were associated with ceremonialism in eastern North America, a thousand years later in Denmark circular earthworks were created to serve as Viking fortifications. From such fortified camps as Trelleborg, longboats set out on their ferocious raids along the North Sea coast and beyond.

During the early nineteenth century a huge earthwork called Lion Mound was constructed at Waterloo to commemorate the defeat of Napoleon's armies in 1815. Another impressive earth monument was erected at Kraków as a lasting memorial to the Polish national hero Tadeusz Kosciuszko, who also served as a general in the American Revolutionary army. After the leader's death in 1817, citizens throughout Poland are said to have brought baskets of earth from their villages to build the memorial.

In the 1950s, twin embankments flanking a football field were built in San Angelo, Texas, to serve as a stadium seating more than twelve thousand spectators. The playing field and running track were excavated 14 feet (4.3 meters) below natural ground level in order to provide fill for the artificial hills.

A pair of artificial hills provides spectator seating for a sunken football field and track in San Angelo, Texas. Courtesy of CRS & Associates Architects, Max D. Lovett, architect.

MOUNDS

Geometrically precise earthworks and a pool create a
serene ensemble known as the Poet's Garden in Zurich,
Switzerland. Photo by Gene Stutz.

One of the most beautiful earthworks of the twentieth century was the
"Poet's Garden" designed by the noted landscape architect Ernst Cramer
for the 1959 Garden Exhibition in Zurich. Consisting of four triangular
pyramids of earth, a conical mound set on a truncated base, and paved
walkways arranged around a rectangular pool, the ensemble is a grass-
encased sculpture garden to be walked through, a serene group of earth
structures designed for contemplation.

Avebury Circle

Wiltshire, England

Older and larger than the nearby earthworks and megaliths of Stonehenge, the multiple rings of Avebury are among the foremost monuments of ancient Europe. Here, on a natural chalk hilltop 90 miles (145 kilometers) west of London, construction began about 2600 B.C. on two circles of stone, one about 340 feet (104 meters) in diameter and the other slightly smaller. Around these two circles a third circle consisting of stones surrounded by a moat and embankment was added perhaps a century later. Workers using stone tools and antler picks excavated and mounded an estimated 200,000 tons of material to create an enclosure measuring 1,396 feet (427 meters) in diameter. At most 30 feet (9.7 meters) deep, the inner ditch may have held water but did not serve defensive purposes.

Avebury's earliest monuments apparently were two large circles of stone erected about 2600 B.C. in Wiltshire, England. Photo by Steve Vidler/SuperStock.

Lined by a deep inner moat, Avebury's outer
circle of earth forms an embankment at most
50 feet (15 meters) high. Photo by Georg Gerster.

Together with the 22-foot (6.7-meter) high embankment, the encircling
wall measured roughly 50 feet (15 meters) in height. Four portals located
at the cardinal points provided entry into Avebury's great circle. Passing
through one of the entrances, visitors long ago may have found before
them a luminous white chalk dome surrounded by a water-filled moat
and a wall of earth enclosing a sacred compound some 28 acres (11 hectares) in area. Possibly one of a series of ancient sites extending for almost
200 miles (360 kilometers) across southern England, the vestiges of the
ancient circle today enclose the town of Avebury.

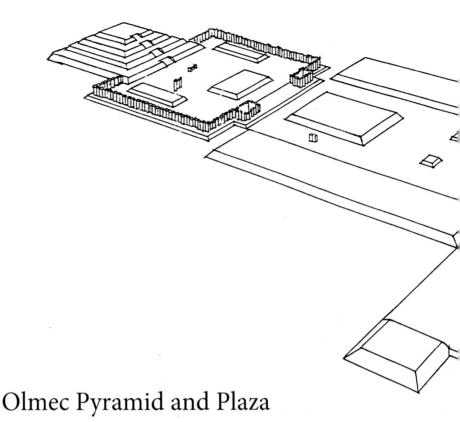

Olmec Pyramid and Plaza

La Venta, Tabasco, Mexico

Predating by several centuries the stone architecture developed elsewhere in Mesoamerica, the earth architecture created by Olmec builders
established design fundamentals that prevailed in the region for the next
2,500 years. These include the creation of large truncated pyramids, the
arrangement of buildings and platforms to form orthogonal plazas, and
the organization of open spaces according to site axes. At La Venta, for
example, Olmec builders laid out a pyramid and plaza complex along a
narrow ridge of clay and sand elevated a few feet above the surrounding
floodplain. Rising some 100 feet (30 meters) above the plaza, the pyramid
was built of clay mixed with sand and small fragments of ceramics; its
basal dimension is 420 feet (128 meters). Probably completed between
800 and 700 B.C., the pyramid now is much eroded and partially covered by an irregular layer of windblown sand up to 8 feet (2.5 meters)
thick. Apparently biaxially symmetrical in plan, the pyramid seems to
have four stairways leading to its small summit, and gently sloping sides
with reentrant corners perhaps along the lines of later Maya buildings

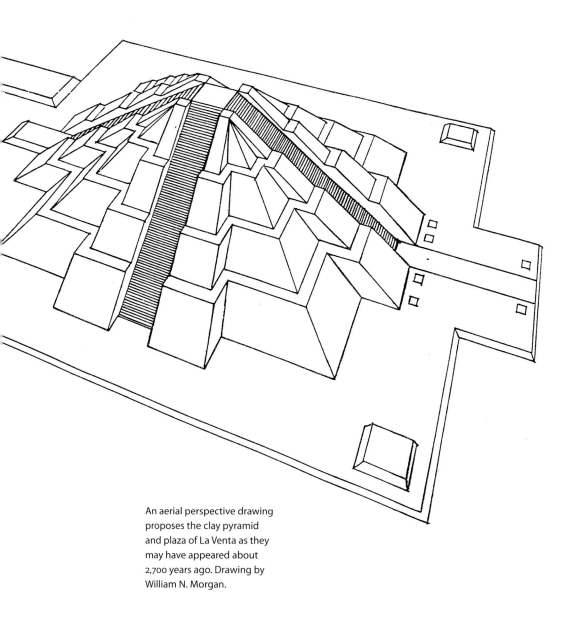

An aerial perspective drawing proposes the clay pyramid and plaza of La Venta as they may have appeared about 2,700 years ago. Drawing by William N. Morgan.

at Uaxactún or Tikal. North of the clay pyramid lies a plaza flanked by earth platforms and a ceremonial forecourt surrounded by a palisade of basalt columns. Earth platforms in the shape of a stepped pyramid form the north terminus of the site axis. Four immense stone sculptures, exotic mosaic pavings, several stelae, and a number of thrones also have been found at La Venta.

Adena Burial Mound

Miamisburg, Ohio

Overlooking the rolling countryside to the north from a 100-foot (30-meter) high bluff near Miamisburg, a conical earth mound rises serenely above the field where it was built some two thousand years ago. Ascending steeply to a height of 70 feet (21 meters) from a base at most 300 feet (90 meters) in diameter, the well-preserved tumulus is the largest Adena burial mound in Ohio. Although amateur investigators excavated a shaft and two tunnels into the mound in 1869, no systematic investigation has been conducted to date. Excavations of similar earthworks elsewhere in the Ohio Valley indicate that the conical earthwork very likely was built by the Adena people, whose identifying characteristics appeared sometime after 500 B.C. and waned early in the Christian era. The Adena are associated with elaborate burial ceremonialism involving specialized burial practices and conical burial mounds. Late Adena tumuli were large and elaborate, often containing spacious burial chambers having one or more bodies accompanied by exotic grave offerings. These include engraved stone tablets, effigy pipes, items made of copper from the Lake Superior area, seashells from the Gulf of Mexico, and mica from North Carolina. Judging by the quality and refinement of grave artifacts, some interments were more important than others, indicating a relatively high degree of social organization. The name Adena is derived from an estate near Chillicothe where in 1901 early investigations of an Adena burial mound were conducted.

Aerial photograph of Miamisburg taken in 1933 by Dache M. Reeves. Courtesy of the Smithsonian Institution National Anthropological Archives, negative 91.

Ceremonial burial mounds of earth were
characteristic of the Adena people who
appeared in the Ohio Valley well more than
2,000 years ago. Photo by Newton Morgan.

Hopewell Octagon and Circle

Newark, Ohio

Native American builders created Newark's immense circular and octagonal earthworks early in the Christian era to serve as one of several ceremonial centers. Photo by Newton Morgan.

Extending more than half a mile (900 meters) across a fertile plain in central Ohio, the geometrically precise circular and octagonal earthworks of Newark are among the best-preserved surviving examples of Hopewell architecture. A grass-covered wall of earth about 6 feet (2 meters) high and 1,050 feet (320 meters) in diameter encloses the great circle that lies southwest of an immense octagon measuring perhaps 1,500 feet (460 meters) across. Eight linear berms define the octagon's perimeter, separated by eight narrow entries. A truncated pyramid inside each entry blocks views into the enclosure; these are defining characteristics of Hopewell architecture. Parallel earth walls set 60 feet (18 meters) apart form an avenue interconnecting the two enclosures. At the southwest terminus of the site axis, the circular walls turn outward for possibly 100 feet (30 meters) as if to form an entry, but this is blocked by a wall of earth some 15 feet (4.6 meters) high. Only by viewing the site from the air or by examining an accurate plan can the site's complex geometry be appreciated. Probably built around 200 A.D., the immense enclosures apparently were related to burial ceremonialism rather than to other human activities. Hopewell ideas about architecture most likely followed their trade networks west into Kansas, south to the Gulf Coast, and northeast into New York and lower Canada. Recorded by Squier and Davis in their historic survey of 1848, the Newark earthworks more recently have been incorporated into a golf course.

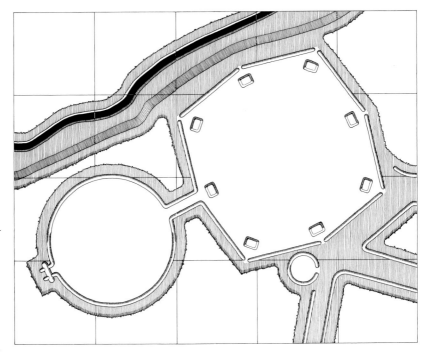

A reconstructed drawing of the Newark site confirms the exceptional scale and precision of the distinctive Hopewell earthworks. The grid scale is 660 feet (200 meters). Drawing by William N. Morgan.

Trelleborg

Slagelse, Denmark

A circular wall of earth almost 600 feet (180 meters)
in diameter protected the ring fort of Trelleborg, one
of several Viking raiding bases in use around 1000 A.D.
Photo by George Gerster.

Originally situated on a cape projecting into shallow water, the Viking ring fort of Trelleborg consisted of a circular stack of turf blocks 565 feet (172 meters) in diameter and at most 10 feet (3 meters) high. Rows of wood posts and tie beams reinforced the earth ring internally. Stone bases and steep walls of wood planks formed the vertical surfaces of the enclosure, and a water-filled moat to the south and east detached the ring fort from the land. At most 40 feet (12 meters) wide, the stepped top of the curving wall consisted of a lower level on the inside and a 3-foot (1-meter) higher level on the outside with a protective breastwork crowning the exterior wall. Four gated tunnels through the turf walls gave access into the central space. Intersecting streets paved with wood planks divided the encampment into precise quadrants; each contained four wood barracks capable of accommodating fifty men without crowding. Built according to a standard plan, the boat-shaped longhouses had column-free interiors and small rooms at both ends. An earthwork southeast of the ring fort enclosed a cemetery and fifteen smaller longhouses that may have been used for storage. Magnificently built and splendidly fortified, Trelleborg was one of four almost identical Viking earth forts that were built quickly, used briefly perhaps between 990 and 1010, and abandoned completely when coastal raids on England subsided. Trelleborg's present-day earth ring identifies the fortification's original position and height but not the details of its construction.

Long ago surrounded by water and projecting into a shallow bay, Trelleborg once was a base for ferocious raids along the English coast. Drawing by William N. Morgan.

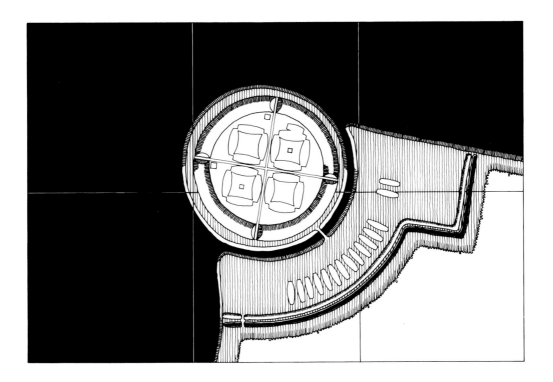

Lion Mound Memorial

Waterloo, Belgium

Leveling the original topography of Waterloo's
battlefield created fill material for the artificial hill
that commemorates the allied armies' victory over
the French. Courtesy of Belgian Consulate General.

From a distance, Lion Mound stands out in the otherwise placid setting where 140,000 combatants once struggled. Courtesy of Belgian National Tourist Office.

Rising more than 140 feet (41 meters) above agricultural fields some 9 miles (14 kilometers) south of Waterloo, an imposing man-made hill of earth commemorates the defeat in June 1815 of the French armies of Napoleon Bonaparte by allied forces commanded by the British Duke of Wellington and Prussian General Blücher. Measuring about 543 feet (166 meters) in diameter at its base, Lion Mound is crowned by the bronze replica of a lion 14.6 feet (4.45 meters) high that is supported by a masonry column descending to original grade through the core of the hill. After its completion in 1826, Wellington called the memorial "a hideous thing" because the original terrain had been destroyed in the process of assembling earth fill to create the great mound. Earth was scraped from the surface of the battlefield, and the 1,300-foot (400-meter) long sunken lane along which Wellington positioned his troops was leveled in the process. Lion Mound marks the site where the Prince of Orange was wounded, and the allied defenses stood firm against repeated assaults by the French cavalry. Today the summit affords unbroken views of the entire battlefield where more than 140,000 combatants once struggled. At the mound's base visitors find a welcome center, souvenir shops, and a round building that houses a panorama depicting a crucial phase of the battle.

Shaped Hills

Greek Hillside Theater, circa 350 B.C., Epidaurus, Greece

Babeldaob, mostly 1000–1400 A.D., Palau, Caroline Islands

Villa d'Este Gardens, begun 1529, Tivoli, Italy

Duke University Stadium, 1928, Durham, North Carolina

Corregidor Bataan Memorial, 1956, Manila Bay, Philippines

Concord Pavilion, Amphitheater, 1975, Concord, California

Low walls stabilize the hillsides and impart a human
scale to the boundless landscape of Honan Province,
China. Courtesy of Japan Press Service.

SHAPING THE earth's surface to serve the purposes of humankind
seems to have been a widespread practice of farmers and builders for
a very long time. One example is the practice of people living in Honan
province of east central China who use low walls to stabilize the surfaces
of mountains created by wind-borne loam. The walls serve not only to
define agricultural plots but also to establish a human scale in the boundless landscape.

On the opposite side of the world, sometime after 400 B.C. Celtic tribesmen began to shape a hill fort southwest of Dorchester, England. Now
known as Maiden Castle, the fort was enlarged over the centuries with
several immense chalk ramparts running around the hillcrest. Maiden
Castle is more notable for the complexity, strength, and beauty of its defensive earthworks than for its size.

SHAPED HILLS

Multiple ramparts successfully defended the Celtic hill
fort of Maiden Castle until the Romans arrived with
siege weapons in 43 A.D. Photo by Georg Gerster.

While Celtic tribesmen were constructing hill forts in northern Europe, Greek architects and artisans were reshaping the natural contours of a gentle hillside at Epidaurus to serve as an amphitheater for thousands of spectators. Now, more than two thousand years later, the remains of the geometrically precise rows of seats are mute reminders of the measured perfection of the Greek ideal closely related to its natural environment.

Many centuries later on a remote island in the western Pacific, one of the world's most spectacular earth sculptures was taking shape. Here, mostly between 1000 and 1400 A.D., the hills of Babeldaob Island in the Palau Archipelago were transformed into a sculpture 27 miles (43 kilometers) long. The reshaped hills still are clearly in evidence today.

At the beginning of the Christian era, Herod the Great reshaped three hilltop earthworks within signaling distance of each other in Palestine. One of these was Herodium, where Herod was buried among spectacular funeral rites. Here the king created an impressive citadel south of Bethlehem that consisted of four towers linked by enclosing walls. The curving stair to the summit was the only means of approaching the fortress.

Created by reshaping a hillside and diverting the waters of a nearby river, the gardens of the Villa d'Este in Tivoli are among the best-known

Curving stairs lead to a reshaped mountaintop containing an ancient fortress and King Herod's tomb, the Herodium. Photo by Georg Gerster.

SHAPED HILLS

Interlocking roads spiral around an urban shopping center in the Helicoide de la Roca Tarpeya, Caracas. Photo by Pedro Neuberger.

examples of formal gardens in sixteenth-century Italy. Terraces, retaining walls, excavation, and fill converted the irregular terrain into a highly ordered, axially symmetrical composition of spirited water features, shaded walkways, gardens, and overlooks.

Distantly recalling the classical Greek amphitheater at Epidaurus, the hillside stadium of Duke University consists of a natural bowl of earth reshaped to receive curving rows of seats. The geometrically disciplined plan affords spectators with unobstructed views of the track and field. Unlike the theater, however, the stadium focuses on a football field instead of a relatively small stage, seats several times as many spectators, and offers distant views no farther than the other side of the stadium.

Responding to a competition for the design of a World War II memorial in the Philippines, the team of sculptor Costantino Nivola proposed to reshape the entire island of Corregidor, the scene of a prolonged siege in 1942. The inspired, but unrealized, design envisioned sculpting the rugged terrain with trenches and tunnels and creating a grove of contemplation with an elevated memorial symbolizing peace and hope.

Strategically located near the center of Caracas, a natural hill known as the Roca Tarpeya was reshaped in the 1960s to serve as a shopping center with a spiraling access road to its perimeter. Integrating architecture with road design, the remarkable project incorporates an access drive that ascends to the summit where an "S" curve begins to descend in an interlocking spiral. The coiling road traverses in all about 2.5 miles (4 kilometers), across the rooftops of some three hundred tenant stores.

Greek Hillside Theater

Epidaurus, Greece

The idyllic countryside and distant mountains provide an ideal background for Greek tragedies and comedies at the theater of Epidaurus. Photo by SuperStock, Inc./SuperStock.

Dramatically situated on a gentle hillside with panoramic views of the distant mountains, the well-preserved remains of the ancient theater of Epidaurus lay hidden under a protective layer of earth until 1881. When construction began about 350 B.C., very little excavation or fill was required because of the site's natural contours. The highly symmetrical plan called for a semicircular bowl of curving rows stepping up the hillside in fifty-five tiers of seats accommodating 14,000 spectators. A curving horizontal walkway separates the lower and upper seating sections. Radiating aisles separate the lower section into twelve segments and the upper section into twenty-two segments. Exquisitely carved marble chairs replace stone benches for seating priests and town dignitaries. All of the seats lie within 200 feet (60 meters) of the center of the perfectly circular orchestra. Measuring 67 feet (20 meters) in diameter, the orchestra was the area for the chorus of each play to dance and sing. Actors performed on the elevated stage of a building behind the orchestra that formed a backdrop for scenery and housed dressing rooms. The original stage building at

SHAPED HILLS

Epidaurus fell into ruins during centuries of neglect. Now designated a World Heritage site by UNESCO, the geometric precision of architectural elements lends dignity and grace to the theater. Classical Greek architecture seems to have maintained its vitality as it evolved, perhaps because the public was sensitive to architecture, required it to be excellent, and took pride in its beauty.

Following the hillside contours, the amphitheater accommodates 14,000 spectators within 200 feet (60 meters) of the orchestra. Photo by age fotostock/SuperStock.

Babeldaob

Palau, Caroline Islands

Sculpted terraces, platforms, pyramids, and domes rise above the jungles of southwestern Babeldaob. Photo by Newton Morgan.

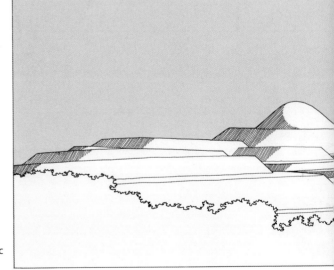

A drawing by the author reconstructs the ancient earthworks from southwest of Ngchemiangel Bay, Babeldaob, an island in the Palau Archipelago of the western Pacific Ocean. Drawing by William N. Morgan.

Sculpted hills seem to extend the entire length of 27-mile (43-kilometer) long Babeldaob, the largest island in the Palau Archipelago. (Babelthuap is an alternate name for Babeldaob.) Among the most impressive views of the extraordinary earthworks are those along Ngchemiangel Bay on the island's west side. The grassy surfaces of the terraces, pyramids, and mounds seem to change colors from red and green to yellow and blue depending on the light, shadow, and time of day. Sculpted from natural hills, the terraced earthworks appear to be part of an island-wide system interrupted at intervals by dense jungles in the valleys and low-lying coastal areas. The crown of one truncated pyramid rises more than 317 feet (97 meters) above sea level and rests on an irregular earth platform measuring at most 200 by 330 feet (60 by 100 meters). Terraces at the bases of some mounds slope inward, perhaps to control erosion or to retain water, suggesting agricultural uses. Stepping down in 15-foot-high intervals, the terraces may have been shaped as needed by small groups of farmers. Occasional ditches and ramps may have provided access to major upland earthworks that probably were built by larger community groups. Uses for the culminating features may have included refuges in times of civil unrest, sites for signal fires, navigational aids, or perhaps symbols of community achievement. Terrace construction on Babeldaob probably began around 500 A.D., intensified between 1000 and 1400, and apparently ceased shortly thereafter.

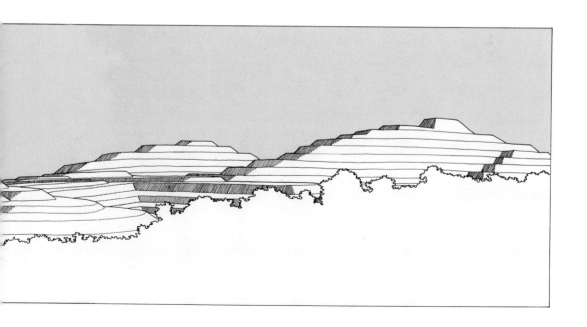

Villa d'Este Gardens

Tivoli, Italy

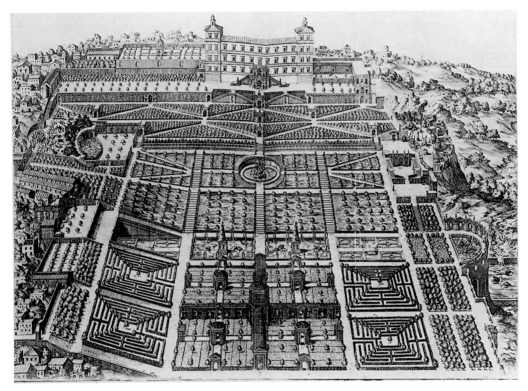

A sixteenth-century print by Duperac shows the palazzo and an early version of the gardens of the Villa d'Este. Courtesy of the Metropolitan Museum of Art, H. B. Dick Fund, 1941.

Enchanting visitors with a sequence of surprises and a sense of mystery, the Villa d'Este probably is the best-known, if not necessarily the best-designed, of all Italian villas. The villa was begun in 1549 for the Cardinal Ipollito d'Este on a site near Hadrian's Villa some 20 miles (32 kilometers) east of Rome. Topographically, the Villa d'Este is located at the crest of a hillside that descends more than 150 feet (46 meters) to the undulating plain below. An engraving dated 1573 illustrates the design's major elements: seven terraces descend from the north-facing palace, bounded by a hill to the east and augmented by a high retaining wall to the west. The site's central axis leads down through the descending terraces; these

SHAPED HILLS

are organized by cross axes, ramps, promenades, stairways, pools, fountains, and an extraordinary array of hydraulic features, supplied by water from the nearby Aniene River. Crossing the central axis, the Terrace of the Hundred Fountains consists of three tiers of sculpted eagles, trefoils, and assorted finials projecting jets and streams of water arching down into troughs. Recessed into the eastern hillside is the flamboyant Water Organ, with a curving gallery behind a thundering waterfall and turbulent reservoirs with towering vertical jets of water recalling organ pipes. Placid pools and shaded lanes on the lowest terrace restore a sense of calm to the gardens now overgrown densely by mature ibex, cypress, and other trees.

Exuberant hillside fountains delight visitors in the upper gardens. Photo by Newton Morgan.

Serene pools and vistas grace the lower terraces of the gardens. Photo by Newton Morgan.

Duke University Stadium

Durham, North Carolina

Created by reshaping a 50-foot (15-meter) deep natural ravine on the campus of Duke University, the distinctive hillside stadium has permanent seating for 35,000 spectators attending football games or track and field events. Built in a remarkably brief time, the impressive stadium began to take shape in November 1928 after trees were cleared and rough grading of the natural bowl progressed. Easily sculpted from well-drained sandy clay soil, the great earthwork measures at most 615 feet (187 meters) in width by 555 feet (170 meters) in length. By May 1929 the football field was leveled, the track was laid out, and concrete was being cast for the lower

Completed only eleven months after clearing of the natural ravine began, the Duke University stadium provides hillside seating for 35,000 spectators. Courtesy of Duke University Archives.

SHAPED HILLS

A view of the ravine in November 1928 records the natural contours as site clearing begins. Courtesy of Duke University Archives.

By May 1929 the modified shape of the ravine clearly resembles a stadium and field. Courtesy of Duke University Archives.

tier of seating. Within five months construction was completed, and the stadium hosted its inaugural football game. Unobstructed site lines were assured by the curving seating plan and the slope of seating rising from the lowest tier 6 feet (2 meters) above the playing field to the highest tier 40 feet (12 meters) above the field. The only Rose Bowl game not played in California was played at Duke in 1942 due to the outbreak of World War II in the Pacific. Supplemented by portable seating, the stadium accommodated a record of 65,000 spectators for the U.S.-Soviet track and field meet in 1974. Now named the Wallace Wade Stadium in honor of a former Duke coach, the enduring earthen bowl continues to serve spectators and athletes today.

Corregidor Bataan Memorial

Manila Bay, Philippines

Strategically situated in the mouth of Manila Bay, Corregidor
Island was proposed to be reshaped extensively to serve as
a World War II memorial. Drawings by Costantino Nivola and
Associates, architects.

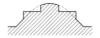

A view of Corregidor Island from South Channel shows the proposed monument elevated above the summit with commemorative features incised into the rugged hillside. Drawings by Costantino Nivola and Associates.

Reshaping 568-foot (173-meter) high Corregidor Island to serve as a World War II memorial was proposed in 1956 by sculptor Costantino Nivola and his associated architects. Located slightly more than 2 miles (3 kilometers) south of Bataan Peninsula in the mouth of Manila Bay, the heavily fortified island withstood months of siege before it capitulated. Developing a concept in which the entire island would become a memorial, the designers envisioned incising, terracing, and tunneling through the rocky terrain to create a series of plazas that would commemorate major campaigns of World War II in the Pacific. At night powerful searchlights would lace the sky with beacons visible from great distances. Visitors would arrive by ferry in the protected small-boat basin on the north side of Corregidor and proceed by way of paths and sometimes sunken passageways to each of the terraces carved into the rugged landscape. Tunnels and trenches would recall the subterranean compartments and gun ports of the original defenses. Arriving on the leveled plateau at the island's summit, visitors would pass through a palm-shaded grove of contemplation with nearby seminar rooms and continue to the sculpted memorial group crowned by an elevated monument symbolizing peace and hope. Elevators in the piers supporting the monument would convey visitors up to a forum level raised above the plateau. The unrealized Corregidor Bataan Memorial would have served as an exceptionally compelling reminder of World War II in the Pacific.

Concord Pavilion

Concord, California

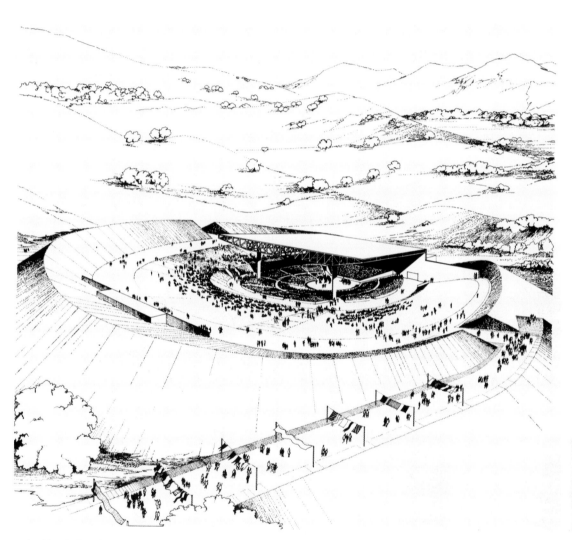

A tilted bowl of earth with a rim some 400 feet
(120 meters) in diameter is reshaped from natural
terrain to serve as the Concord Pavilion. Courtesy
of Frank O. Gehry, architect.

Reshaped from a natural hillside to become an amphitheater, the Concord Pavilion lies in the rolling terrain of the Diablo Range about 25 miles (40 kilometers) east of San Francisco. Rising as high as 60 feet (18 meters) above the stage, the surrounding berm deflects wind and traffic noises from the northwest while improving acoustics in the earthen bowl. Visitors leave their vehicles near the highway and ascend by way of a gently curving ramp to the entry on the berm's crest. From this point they observe a circular seating area partly covered by a roof measuring 200 feet (60 meters) square and surrounded by grass-covered slopes accommodating a total of 8,000 viewers. A wall extending across the south side of the pavilion serves as the backdrop for the stage while concealing from view the loading dock and backstage functions. Two concrete columns supporting the steel roof trusses are carefully located for minimum interference with sight lines. Adaptable to theater-in-the-round, proscenium, or thrust-stage configurations, the circular stage lends itself to a wide variety of presentations. Catwalks suspended from the steel trusses provide access to the control booth, lighting devices, and acoustical equipment. Designed in the early 1970s by architect Frank O. Gehry and his associates, the Concord Pavilion capitalizes on the moderate local climate and attractive hill country setting. Although an estimated 250,000 cubic yards (190,000 cubic meters) of earth were moved to complete the project, the character of the site remains essentially intact.

Only the grassy berm of the pavilion can be seen from the parking area. Photo by Morley Baer.

 # Earth Retained

Machu Picchu, Late Inca Town, 1450–1550, Urubamba Valley, Peru

Second Jacobs House, 1948, Middleton, Wisconsin

Expressways, late twentieth century, United States

New Cemetery Proposal, 1973, Urbino, Italy

Vietnam Veterans Memorial, 1982, Washington, D.C.

Spencertown House, 2003, Hudson Valley, New York

RETAINING EARTH requires a wall or other structure to hold the earth in place above or below the level of adjacent ground. Illustrating the principle, Etruscan builders used stone walls to retain earth mounds above grave chambers called tumuli, and to hold back the ground above subterranean streets in cities of the dead such as the necropolis of Cerveteri. Niches or alcoves sometimes were carved into the walls to serve as burial places. Visitors to the necropolis may wander through the irregular streets without seeing the horizon in the distance; they see only the sky.

Inca builders used retaining walls to create Machu Picchu and other cities on spectacular sites high in the Andes Mountains of Peru. Closely following natural contours, walls of carefully fitted granite stone retained earth fill for linear gardens, building foundations, and community plazas. Nearby Ollantaytambo probably is the best-preserved Inca town, with its well-conceived site plan, sensitive integration of structures with the natural environment, the unity of its architecture, and the precision of its fitted stone masonry.

In more recent times, Frank Lloyd Wright employed a curving stone wall two stories high to serve as a wind deflector and structural enclosure for the Second Jacobs House in Wisconsin. The building curves around

Retaining walls encircle earth mounds above tumuli that line subterranean streets in the Etruscan necropolis of Cerveteri, Italy. Photo by William N. Morgan.

EARTH RETAINED

Precisely fitted granite stones retain earth structures integrated with the mountainous terrain in the late Inca town of Ollantaytambo, Peru. Photo by Ludwig Glaeser.

a garden that is sunken into the earth to the south, relying on its massive north wall for insulation during the winter and for cooling during the summer.

Often utilizing retaining walls for bridge approaches and overpasses, highway construction in the United States each year involves the relocation of immense quantities of earth. Tollgates, cloverleafs, multilevel interchanges, and holding ponds are reshaping our cities and countryside with rigorous efficiency, but too often at the expense of our environment. Balancing ecological preservation with transportation needs suggests new directions in urban and regional design.

Designing a new capitol of Punjab in 1952, Le Corbusier envisioned an arrangement of reflecting pools on several levels recessed into the forecourt of the Governor's Palace at Chandigarh. Reflections on different levels were intended to avoid the illusion of great distances between the palace and its nearby governmental buildings. Retaining walls held back the ground above the sunken courtyards with their precisely proportioned stairways, sculptures, pools, loggias, and pavings.

Walls retaining landscaped earth berms surround the perimeter classrooms and courtyard of the Patton Elementary School on a hillside in Monterey County, California. Enclosed on all sides by a covered walkway, the central quadrangle is excavated into the site to accommodate play spaces and ancillary buildings. The berm-encased buildings and courtyard provide a strong sense of place among the windswept sand dunes not far from the Pacific coast.

Distantly recalling the Etruscan necropolis of Cerveteri, the proposed new cemetery for Urbino, Italy, consists of walled subterranean streets lined with tombs and recessed into a rounded hilltop adjoining the old

Retaining walls below grade define richly varied spaces that are enhanced by pools, stairways, and sculptures in the forecourt of the Governor's Palace at Chandigarh, India. Drawing by Le Corbusier.

EARTH RETAINED

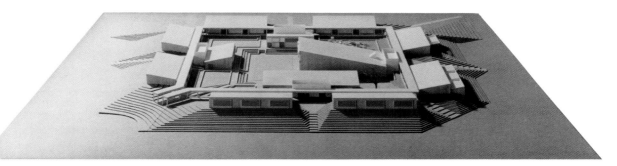

Landscaped berms protect an elementary school's
classrooms and courtyard from coastal winds on
an exposed hillside in Monterey County, California.
Courtesy of Wallace Holm, architect, and associates.

cemetery. Exploring the theme of outer geometric perfection in contrast
to inner chaos, the width of the street between the tombs varies unpre-
dictably, and the spaces around the sepulchers constantly change.

Synthesizing architecture, landscape architecture, and sculpture, the
Vietnam Veterans Memorial consists of a polished black granite retaining wall that increases in height as visitors descend into a special place of
contemplation recessed into the earth. Veterans' names are inscribed into
the mirror-like surface, recording the war as a series of human sacrifices.
Broken into two sections at its lowest point, the wall points to the Washington Monument in one direction and to the Lincoln Memorial in the
other.

A reinforced concrete retaining wall more than 100 feet (30 meters)
long extends across an open lawn that slopes gently down to the west
near Spencertown, New York. The retaining wall serves as the east wall of
a one-story-high residence that responds to nature without trying to look
like nature. The design reads as a single linear volume in the landscape,
advancing environmentalism by staying low to the ground and providing
the house with shade and breezes.

Machu Picchu

Urubamba Valley, Peru

Spectacularly sited on a saddle-shaped ridge in the rugged Peruvian Andes, the late Inca town of Machu Picchu lies between 10,500-foot (3,200-meter) high Mount Machu Picchu to the south and a lower peak to the north. Elevated some 2,000 feet (600 meters) above the winding valley of the sinuous Urubamba River, the town consists of buildings and plazas to the north, with mountainside gardens to the south. Retaining walls built of carefully fitted granite stones support the linear gardens following the contours of the steeply sloping terrain. The intensely cultivated terraces were filled with fertile soil brought up from the valley below probably on the backs of llamas, the Inca's only draft animal. A mile-long (1.6-kilometer) aqueduct irrigated the agricultural plots and conveyed water to a

Fertile soil brought up from the valley below filled the numerous terraces of self-sufficient Machu Picchu. Photo by age fotostock/SuperStock.

EARTH RETAINED

Well-constructed buildings with impressive stone walls flank the lowest terrace of Machu Picchu's central plaza. Photo by Digital Vision Ltd./Super-Stock.

street with sixteen descending fountains near the town's center. Probably planned and built under the supervision of professional Inca architects, self-sufficient Machu Picchu is laid out on both sides of several spacious plazas stepping gently down along the central axis. The sun platform and sacred precinct with the main temple and courtyard line the western ridge; to the east lie the cemetery, mausoleum, and clan houses. The palace and house blocks for commoners are situated between the plazas and the agricultural zone. Located about 43 miles (70 kilometers) northwest of Cuzco, well-preserved Machu Picchu probably was built toward the end of the Inca hegemony between 1442 and 1532 A.D.

Second Jacobs House

Middleton, Wisconsin

A stone-paved walkway leads visitors through a
tunnel in the bermed north wall of the Second
Jacobs House. Photo by Ezra Stoller.

EARTH RETAINED

The curving north wall rises two stories above the seemingly endless midwestern prairie. Photo by Ezra Stoller.

Shaped like a grass-covered hill on the edge of the seemingly endless prairie, the curving north wall of the Second Jacobs House is a two-story-high earth berm retained by a stout stone wall and buttressed by a circular tower. One of several well-resolved designs by Frank Lloyd Wright that use earth as a building material, the project is the second house that he designed for his journalist friend Herbert Jacobs. Also known as the "Solar Hemicycle," the building is organized around a sunken garden facing south with a 48-foot (15-meter) long curving glass wall along its north side. Lying between the earth-bermed north wall and the circular bowl to the south, the design develops an airfoil that deflects cold winter winds. The 4-foot (1.2-meter) wide roof overhang protects the south-facing glass wall from the sun in the summer but admits solar gain deep into the interiors during the winter. In the summer the stone walls naturally cool the interior spaces, while high windows between the berm crest and roof overhang vent warm air to the north. Visitors enter the house by way of a stone-paved walkway that leads through a tunnel near the berm's east end. Viewing the sunken garden as they approach, visitors turn to the right and enter through a door in the glass wall into the kitchen and dining area. Beyond lie the circular fireplace recessed into the north wall and a round pool half inside and half outside of the living room. Stairs curve up the tower walls to the bedrooms and bath above on the mezzanine level.

Expressways

United States

Growing emphasis on environmental preservation
and pollution control increases the complexity of
highway design. Photo by age fotostock/SuperStock.

Expressways are reshaping our environment and demanding far-sighted, creative, and imaginative designs. Photo by Alex MacLean.

Expressways are primary determinants of urban and regional developments in the United States today. Reshaping the environment to accommodate contemporary highway systems often involves the relocation of immense quantities of earth, requiring extraordinary new earth-moving machines. The reshaping process presents unprecedented opportunities to enhance, preserve, or destroy our cities and countryside. Contemporary architecture is being called upon to create farsighted designs that equal or exceed the rigorous efficiency and boldness of contemporary engineering. Disregarding environmental, ecological, or cultural considerations in the past, road construction sometimes has flattened hills and valleys, sliced through mountains, destroyed wetlands with fill, and fragmented traditional urban neighborhoods. Tollgates, cloverleafs, multilevel interchanges, and holding ponds consume large parcels of land while increasing the complexity and expense of expressways. Growing emphasis on environmental protection, pollution control, and ecological preservation encourages urban and regional designs that are sensitive to the needs of human beings and their surroundings. Rethinking our transportation systems on a fundamental level may reorder our priorities and suggest new directions for creativity and imagination. Consideration may be given to routing arterial roads around neighborhoods or urban centers that would both preserve and enhance them, and devising alternative light rail or waterway systems that would reduce dependence on roadways.

New Cemetery Proposal

Urbino, Italy

Subterranean streets flanked by irregularly spaced tombs are incised into a gently rolling hilltop in Urbino, subtly contrasting disarray with perfection. Photo by Arnaldo Pomodoro.

Responding to a design competition in 1973, sculptor Arnaldo Pomodoro contrasted the perfection of a gently curving hilltop with the imperfection of eroded streets cut deeply into the earth. Distantly recalling an ancient Etruscan necropolis with its main funeral road open to the sky above and flanked by tombs below grade, Pomodoro's design explores a timeless metaphor in contemporary art and architecture. A new wall marks the boundary between the proposed project and the old cemetery, with its traditional and predictable disposition of elements. An opening in the wall provides an entry into a subterranean passageway flanked by

EARTH RETAINED

tombs arranged so that the width of the street and the spaces around the sepulchers constantly change. The entry lane leads to the main funeral road that continues to the right and left. A second lane and narrow slits at unexpected intervals turn into the hillside, while a short street leads to an open space below grade that surrounds a square tomb. At every opportunity the design returns to the theme of outer geometric perfection reduced to inner chaos. The streets terminate in eroded paths that ascend by means of gently sloping ramps to the curving hillside above where order, serenity, predictability, and a sense of perfection prevail once more. The design serves not only as a sculptural work of art but also as a functioning component in the daily life of the community. The subject of enormous controversy, the design won the competition but was rejected by the authorities.

The proposed new cemetery clearly recalls an Etruscan necropolis built elsewhere in Italy long ago. Photo by Arnaldo Pomodoro.

Vietnam Veterans Memorial

Washington, D.C.

At the Vietnam Veterans Memorial, a ramp gently descends into a special place of contemplation recessed into Constitution Gardens. Courtesy of Maya Lin Studio.

Created by sensitively reshaping the terrain of Constitution Gardens, the Vietnam Veterans Memorial consists of a polished black granite retaining wall that gradually increases in height as visitors descend into a special place of contemplation recessed into the earth. Here the sights and sounds of the surrounding city no longer distract visitors. Carved into the wall's mirror-like surface, veterans' names are inscribed according to the day of casualty, recording the war as a series of human sacrifices. At the ramp's lowest point of descent, the wall breaks at an obtuse angle; one section of the wall points toward the distant Washington Monument, while the other points to the nearby Lincoln Memorial. The wall's orientation locates the Vietnam Veterans Memorial symbolically in the nation's history, while

EARTH RETAINED

The wall reflects the images of present-day viewers and the surrounding landscape. Courtesy of Maya Lin Studio.

Casualties' names are inscribed into the memorial's polished marble walls. Courtesy of Maya Lin Studio.

its polished surfaces reflect the images of present-day viewers and the surrounding landscape. Serving as a monument to heal a nation, the brilliant design is the clear vision of Maya Lin, who was an architectural student at Yale when she created her competition-winning proposal. A synthesis of architecture, landscape, and sculpture, the Vietnam Veterans Memorial is one of the most compelling works of architecture to be created in the twentieth century.

Spencertown House

Hudson Valley, New York

Taking maximum advantage of an open 14-acre (6-hectare) site rolling gently down to the west, architect Thomas Phifer's design for the Spencertown house responds to nature without trying to look like nature. Treeless except at its perimeter, the site has a small plateau near its center where the narrow, linear building is situated. A reinforced concrete wall more than 100 feet (30 meters) long extends across the plateau and forms the east wall of the residence, retaining the higher terrain of the entry lawn to the north. Operable windows above the retaining wall and openings in the glass wall to the west provide through-ventilation when required. Recessing the floor slab and east wall into the earth takes maximum

A curving metal roof hovers above an earth berm
more than 100 feet (30 meters) long that organizes
the Spencertown House along a north-to-south axis.
Photo by Scott Frances.

The extensive overhang and continuous sunscreen shade the glazed west elevation with its spectacular views of the distant mountains. Photo by Scott Frances.

advantage of the earth's year-round temperature of 55 degrees Fahrenheit (13 degrees Celsius). Approaching from the east, visitors observe the curving metal roof hovering above the continuous transom that separates the eave from the landscaped terrain. An entry court slightly north of the building's center separates the guesthouse to the right from the residence proper to the left. Uninterrupted glass walls to the west present spectacular views of the Catskill Mountains in the distance beyond the forest. An extensive roof overhang and continuous sunscreen in the upper wall plane reduce heat gain and glare late in the day through the glazed west wall of the residence. Reading as a single linear volume in the landscape, the design advances environmentalism by staying low to the ground and providing shade with breezes.

Terraces

Hatshetsup's Mortuary Temple, 1520 B.C., Deir el Bahri, Egypt

Governor's House, 850–915, Uxmal, Yucatán, Mexico

Ifugao Rice Terraces, traditional, Luzon, Philippines

Maori Forts, 1500–1800, North Island, New Zealand

Middleton Place, 1741, Charleston, South Carolina

Marae of Mahaiatea, circa 1760, Tahiti, Society Islands

Orderly rows of wheat terraces transform the precipitous terrain of Shikoku Island into productive agricultural fields. Photo by Burt Glinn.

OVER THE CENTURIES, terraces have been used not only for agriculture and housing but also as components of impressive monuments with dramatic results. For instance, Queen Hatshetsup's mortuary temple is placed on ascending terraces at the base of sheer cliffs along the Nile Valley's western edge. Another example of monumental buildings closely associated with terraces is the House of the Governors at Uxmal, where four levels serve an impressive complex that functions both as a formal residence and as a governmental center.

Among the oldest and most extensive agricultural terraces in this study are the ancient rice terraces built by the peaceful and industrious Ifugao farmers who inhabit the mountainous interior of Luzon Island in the northern Philippines. In contrast to the peaceful and constructive enterprises of the Ifugao farmers, the Maori people who lived far to the south in prehistoric New Zealand created thousands of terraced earth forts

known as *pa* for the storage and protection of the food they produced from hungry and bellicose neighbors.

About the time that *pa* building was reaching its peak in New Zealand, terrace building attained a new level of sophistication in North America with the construction of the grass-covered terraces, tidal ponds, and causeways at Middleton Place in South Carolina. The builders were Low Country rice producers who thoroughly understood the art of terrace building.

Meanwhile, on the opposite side of the world, people living on Tahiti and neighboring islands in the South Pacific were developing a unique type of terrace called *marae*. Prominent families competed with each other by building multilevel terraces that resembled elongated pyramids. Higher *marae* represented higher social and political status on some islands of Oceania.

Derived from the Latin word meaning *earth*, terraces are level surfaces of earth with sloping or vertical sides rising one above the other. Clear examples of terraces are the wheat fields stepping up rugged mountainsides on Shikoku Island, the fourth largest of Japan's home islands. In many sections of the landscape, orderly rows of wheat terraces have replaced former forests, converting the precipitous terrain into highly productive agricultural land. Tidy stacks of wheat appear where the crops have been harvested, occasionally interrupted by rows of trees or patches of forests.

Other examples of agricultural terraces are those of San Felices, Soria, in mountainous terrain that lies northeast of Madrid. Centuries ago, industrious farmers here converted barren hillsides into productive fields by

Ancient Spanish hillside terraces continue to yield food crops annually in San Felices, Soria. Courtesy of Spanish Travel Bureau.

constructing terraces faced with fieldstones that were collected from the rocky soil. Closely following the natural contours and irrigated by small springs, the terraces contain fertile soils that yield a variety of food crops annually.

During the 1950s terraces were designed for a compact community of townhouses on a gentle slope overlooking a forested valley near Berne, Switzerland. The terraces step up in carefully designed increments that assure the preservation of the environment and views of the river valley, including the nearby Halen Bridge. Known as Siedlung Halen, the hous-

The compact Halen Community housing preserves privacy as well as its hillside environment near Berne, Switzerland. Photo by Albert Winkler.

The low-density housing terraces of
Trousdale Estates consume a suburban
mountainside in Los Angeles, California.
Photo by Pacific Air Industries.

ing group contains residences with a high degree of privacy. The dwellings
are individually owned and exceptionally energy efficient.

At the same time that Siedlung Halen was being constructed, a housing
project known as the Trousdale Estates was being developed on an extensively terraced mountainside north of Los Angeles. The California project
called for low-density single-family detached dwellings on individual terraces connected by steep access roads and extensive utility lines. The forested hillside disappeared and privacy in each residence was diminished,
the opposite of the experience in Switzerland.

Hatshetsup's Mortuary Temple

Deir el Bahri, Egypt

At Hatshetsup's Mortuary Temple, an avenue flanked by sphinxes leads to colonnades and ascending terraces in a spectacular natural setting. Photo by Newton Morgan.

Dramatically situated against a backdrop of sheer cliffs along the western edge of the Nile Valley, the mortuary temple of Hatshetsup is justly celebrated as an excellent example of harmony between architecture and nature. Consisting of three magnificent earth terraces stepping up as they approach the base of the mountains, the temple is approached along an avenue of sphinxes that leads from the plain to a gateway forming the entrance to the lowest terrace. Handsomely proportioned colonnades retain the east walls of the serene upper terraces that are interconnected

TERRACES

Stoutly buttressed walls define the terrace perimeters while the central sanctuary is carved deeply into the mountainside. Photo by Newton Morgan.

by broad, gently ascending ramps. The ramps are symmetrically aligned along the centerline of the site's east-west axis, but the second terrace is shifted to the north and requires a stoutly buttressed retaining wall to the south and a walled colonnade on the north to hold back the earth. The colonnades of both levels align visually one above the other, but an additional mass, the Chapel of Anulis, is added to the north to account for the terrace offset. The Chapel of Hathor balances the composition to the south, but it must be entered from the south colonnade because it lacks a terrace. The upper terrace returns to the temple's symmetrical composition with its flanking court and chapel as well as a central sanctuary carved deeply into the mountain. The subtly resolved composition of axially shifting elements is attributed to the temple's noted architect, Senmut, who aligned the ensemble with the temple complex of Mentuhotep built five centuries earlier to the south.

Governor's House

Uxmal, Yucatán, Mexico

Four distinctive terraces set the House of the Governor apart from its monumental neighbors. Photo by Newton Morgan.

Rising as high as 50 feet (15 meters) above the surrounding plain, the four grand terraces of the Governor's House are no less impressive than the much acclaimed building that they support. Recorded with a surprising degree of accuracy in Frederick Catherwood's 1843 engraving, the terraces required perhaps four times as much time to construct as the edifice proper. The lowest terrace measures 625 feet (190 meters) along the entry side by 570 feet (174 meters) along the north side, and rises 3.5 feet (1 meter) or so above the irregular terrain. Rounded limestone blocks define the terrace's corners; stone walls between the corners retain earth and rubble fill that is surfaced with graded pebbles and likely paved with lime mortar mixed with naturally occurring white earth. The second terrace is 20 feet (6 meters) higher than the first; it forms a broad public plaza with a distinctive jaguar throne axially centered on the building. Measuring 540 by 500 feet (165 by 152 meters), the plaza incorporates the House of the Turtles to the northwest and abuts the Great Pyramid to the southwest.

TERRACES

A broad stairway leads 21 feet (6.4 meters) up to the third terrace that is retained by four-stepped walls. A short stair run leads up to the fourth terrace at the floor level of the 30-foot (9-meter) high building. Probably completed shortly after 900 A.D., the House of the Governor apparently served both as an elite residence and as an administrative center with an impressive plaza and chambers suitable for formal audiences.

The 21-foot (6.4-meter) high third terrace establishes an impressive base for the building's modest end elevation. Photo by Newton Morgan.

Ifugao Rice Terraces

Luzon, Philippines

Rivers high in the mountains above Ifugao irrigate
rice terraces chiseled into steeply sloping hillsides.
Photo by SuperStock, Inc./SuperStock.

Raised on stilts near the rice fields, houses often exhibit fine craftsmanship. Photo by Robert Höbel.

In the rugged mountains of northern Luzon, Ifugao farmers have been converting precipitous slopes into rice terraces for thousands of years. Here the area of agricultural terraces far exceeds that of China, Southeast Asia, Java, Japan, or elsewhere on earth. The peaceful and industrious Ifugao people are at the center of terrace building activity. Depending on the source of water for irrigation, rice fields are constructed as high as 4,265 feet (1,300 meters) above sea level near the town of Banaue, where mountain peaks rise well over a mile high. Generous rainfall assures numerous streams and springs fed by reservoirs in the volcanic rock. Terrace walls of stone, shale, or mud typically range from 6 to 33 feet (2 to 10 meters) in height. Construction implements include primitive crowbars, hammers, and spades often made of iron, using a technology known in the region for several thousand years. Ifugao communities consist of 5 to 60 families under the leadership of their agricultural chief, a descendant of the first terrace constructors, who typically inherited the widest part of a nearby river valley. Here maximum areas of terraces can be created with minimum effort. Canals, trenches, spillways, and bamboo tubes convey irrigation water across the terraced mountainsides. During fallow periods the rice fields remain flooded to keep the soil moist, control landslides, reduce weeds, and facilitate replanting. Built on stilts near the rice fields, Ifugao houses often exhibit superlative wood frame craftsmanship.

Maori Forts

North Island, New Zealand

Such Maori earth-
works as those of
Omuna may have
served not only as
forts but also as sta-
tus symbols and food
repositories. Photo
by Nigel Prickett.

Creating earthworks associated with as many as 6,000 fortified places called *pa* was the most labor-intensive activity undertaken by the Maori people of prehistoric New Zealand. In the centuries following the arrival of Polynesian voyagers around 800 A.D., populations increased steadily in relation to the successful production of sweet potatoes, particularly on North Island. As populations soared, competition for productive agricultural lands led to a settlement pattern dominated by forts built in naturally defensible locations. By 1500 A.D. defensive ditches and palisades were

TERRACES

incorporated into *pa* of widely varying scales and configurations. For example, the ring fort of Omuna lies along the coast of the Tasman Sea a short distance south of New Plymouth. A gully to the north and another to the south, and a steep slope down 80 to 100 feet (25 to 30 meters) to the beach protect the *pa* on three sides. The encircling ditch and bank close off the fourth side. Rising 12 to 20 feet (3.5 to 6 meters) above the surrounding ditch, the central platform measures at most 50 by 60 feet (15 by 18 meters). Two collapsed food storage pits are the platform's only features. Other examples of Maori earthworks are the enormous and spectacular artificial terraces created by reshaping the eastern slope of Mount Willington, one of Auckland's several volcanic cones. Fortified settlements of this type typically contained provisions for dwellings, facilities for food processing and storage, cooking areas, and other activity spaces.

Maori builders invested enormous energy in creating the spectacular terraces on the eastern slope of Mount Wellington in Auckland. Photo by Janet Davidson.

Middleton Place

Charleston, South Carolina

Earth-shaping techniques developed for rice cultivation
evolved into lakes, esplanades, and terraces at Middleton
Place in the 1700s. Courtesy of Middleton Place.

TERRACES

Tidal marshes and cypress swamps along the Ashley River were transformed into terraced gardens complementing the residence halls of Middleton Place. Courtesy of Middleton Place.

The grass-covered terraces of Middleton Place step gently down to the level of the Ashley River some 15 miles (24 kilometers) southwest of Charleston, South Carolina. From the base of the terraces an earth causeway 50 feet (15 meters) wide leads toward the riverbank between two lakes shaped like butterfly wings. The formal design was laid out in 1741 and was largely completed within a decade by sculpting the 25-foot (8-meter) high natural embankment into five equal earth terraces, each 30 feet (9 meters) wide. The 350-foot (107-meter) wide lawn overlooking the terraces was the site of the original plantation house, of which only the south wing remains today. The plantation's economy was based largely upon growing rice in the tidal ponds during the eighteenth and nineteenth centuries. Earth dikes and floodgates controlled the water level as the river level fluctuated with tides. Coastal ponds were preferred for rice cultivation because tree clearing was not required, and the water supply was plentiful. Over the years extensive gardens were developed to the north and west of the mansion house, where camellias, azaleas, and magnolias thrive still. Today the well-maintained terraces of Middleton Place continue to be among the most prominent features of the oldest extant landscaped gardens in North America.

Marae of Mahaiatea

Tahiti, Society Islands

Ten impressive terraces of diminishing sizes once formed a stepped pyramid of earth and stone that rose 50 feet (15 meters) above the sandy south shore of Tahiti. According to Captain Cook, who observed the structure in 1769, a year or two after its completion, the elongated pyramid measured about 87 by 267 feet (27 by 81 meters) at its base and was situated toward the west end of a stone-paved terrace surrounded by a low stone wall. Elevated several feet above the adjacent grade, the terrace extended the full width of the pyramid for perhaps 377 feet (115 meters) parallel to the shore. The terrace and pyramid together constitute a religious structure known as a *marae*. Near the middle of the pyramid's summit Captain Cook found a bird figure carved of wood and a fish figure sculpted in stone, suggesting the structure's symbolic function as an altar. Stones erected in front of the pyramid served as seats for the gods or for the directors of the ceremonies. In the middle of the terrace were tables on which offerings to the gods were displayed, and seats consisting of stones carved with motifs indicating the social rank of the attendants. An important family erected the *marae* of Mahaiatea, the largest in Tahiti, for their

Dressed basalt corner stones and water-worn cobbles line the terrace walls of the restored four-step Marae of Ahu O Maline on Moorea Island. Courtesy of Bernice P. Bishop Museum Archives.

The exceptional height of the Marae of Mahaiatea
on Tahiti symbolized the elevated status of the
family that erected it. Courtesy of Bernice P. Bishop
Museum Archives.

son who was heir to the district high chief. Although the structure now
is much in ruins, the four-step *marae* known as Ahu O Maline on neighboring Moorea Island has been restored using such authentic materials
as dressed basalt or coral bases, vertical corner stones with in-filling, and
water-worn cobble facings.

 # Platforms

Persepolis, 518–330 B.C., Shiraz, Iran

Borobudur, Buddhist Stupa, 778–842 A.D., Java, Indonesia

Preah Vihear, Buddhist Temple, 800–1000, Cambodia

Emerald Mound, 1400–1600, Natchez, Mississippi

Governor's Icehouse, 1700s, Williamsburg, Virginia

Dallas–Fort Worth Airport, begun in 1971, Grapevine, Texas

Elevated 1,500 feet (460 meters) above the Valley of Oaxaca, the magnificent Zapotec ceremonial center of Monte Alban was created by leveling a mountaintop. Courtesy of Instituto Nacional de Antropología e Historia.

PLATFORMS ARE horizontal planes that often are higher than their adjoining surfaces. One of the most spectacular platforms in this study is the lofty plateau of Monte Alban, the noted Zapotec ceremonial center elevated some 1,500 feet (460 meters) above the Valley of Oaxaca in central Mexico. Built mostly between 200 and 900 A.D., the grand platform incorporates natural rock outcroppings and the ruins of earlier structures into the truncated pyramids, tombs, ballcourt, and walled courtyards located in and around the central plaza. Regarding Monte Alban, Aldous Huxley (1934) commented, "few architects have had such a sense of austerely dramatic grandeur.... Religious considerations were never allowed to interfere with the realization of a grand architectural scheme."

Another platform of "austerely dramatic grandeur" is the Altar of Heaven built in the center of the Chinese Empire in 1420. The sacred place consists of three perfectly circular stone-faced platforms of diminishing diameters set one on top of the other and surrounded by a circular wall. Portals and stairways at the cardinal points give access to the center of the top platform, where a circular slab marks the Circular Mound Altar. No structure is placed above the altar so that men can talk directly to the superior being.

Two college libraries built recently in the United States illustrate different types of platforms that carefully integrate architecture with campus planning. Near the center of the Hendrix College campus in Conway, Arkansas, a large platform has been excavated several feet below grade to serve as a forecourt for berm-encased Bailey Library. Visitors to the library descend from surrounding walkways and courtyards to the level of the spacious plaza and enter directly into the building. This approach

Four portals and stairways lead up to the Altar of Heaven in the center of three perfectly round platforms surrounded by a circular wall in Beijing.

The forecourt of Bailey Library is a large entry plaza excavated into the earth of Hendrix College in Conway, Arkansas. Courtesy of Philip Johnson, architect.

gives visitors the impression of a level entrance into the library rather than a descent into the ground.

Closely related to existing buildings in Harvard Yard, the rooftop of the new Pusey Library serves as a pedestrian thoroughfare. Elevated at most 9 feet (2.7 meters) above natural grade, the platform occupies a strategic location along the heavily traveled pedestrian walkway at the southeast corner of Harvard Yard. The landscaped structure's west edge serves as a broad canopy above the glazed entrance into the library.

Much older and much larger than the foregoing platforms, the impressive podium of ancient Persepolis overlooks a broad plain in southern Iran. The imperial city once was the center of ancient Persia; its ruins today recall the magnificence and splendor of the grand empire. A different, more spiritual sense of magnificence is conveyed by the spectacular pilgrimage shrine of Borobudur in Java; here nine platforms of diminishing sizes are ornately embellished and sculpted to symbolize the Mount Mehru of Buddhist mythology.

PLATFORMS

Another compelling example of pilgrimage shrines in southeast Asia is the extraordinary Preah Vihear temple complex in Cambodia, where ascending platforms create a sacred processional way that culminates above a dramatic cliff. Less architecturally sophisticated but no less spiritually rewarding is the Native American ceremonial center known as Emerald Mound, an enormous platform of earth crowned by truncated pyramids and a central plaza near the south terminus of the Natchez Trace in Mississippi.

Truncated platform mounds superimposed one upon the other contain the well-insulated icehouse of the eighteenth-century Governor's Palace in Williamsburg, Virginia. Surmounted by an observation platform, the structure anchors the northwest corner of the colonial governor's extensive gardens, earthworks, and terraces. Another utilitarian platform in this study and perhaps the largest in terms of the quantity of earth relocated is the Dallas–Fort Worth airport in Texas. The recently completed project required a level platform almost four miles (6.4 kilometers) long to be created across rolling terrain, gullies, and dry washes—a daunting task even with twentieth-century technologies at the builders' disposal.

Situated in the southeast corner of Harvard Yard in Cambridge, Massachusetts, Pusey Library's rooftop preserves the spaces below and links the pedestrian walkway above. Courtesy of Hugh Stubbins Associates, architects.

Persepolis

Shiraz, Iran

A vast earth platform some 40 feet (12 meters) high elevated the magnificent ceremonial city of Persepolis above the eastern plain. Photo by Georg Gerster.

Elevated on an impressive platform of earth and stone some 40 feet (12 meters) above a broad plain in southern Iran, the magnificent palace complex of Persepolis was founded by Darius the Great about 518 B.C. Located at the base of the Mountain of Mercy some 30 miles (48 kilometers) northeast of present-day Shiraz, the vast platform measures about 1,000 by 1,600 feet (300 by 500 meters) in size. Visitors enter the imperial city by way of the richly decorated monumental stairs at the northwest corner of the platform. On the summit now lie the ruins of colonnaded audience halls, palaces, courtyards, gatehouses, the treasury, and harem. Overlooking the entry court, the Apadana Hall rests on a leveled stone bluff some 20 feet (6 meters) above the level of the platform. Building walls typi-

Staircases embellished with stone carvings guided delegations up to the main podium level and to the audience hall above. Photo by Ludwig Glaeser.

cally consisted of sun-dried and fired clay bricks with stone foundations and portals. Glazed tiles or stucco with colorful designs faced the lower portions of some walls. Channels built into the structures and platform conveyed water to a large stone reservoir under the paving. Disproving speculation that slave laborers erected Persepolis, excavators have found numerous clay tablets recording substantial wages paid in silver to construction workers. In 330 B.C. Alexander the Great destroyed Persepolis and, according to Plutarch, carried away the treasures of the empire on 20,000 mules and 5,000 camels. Today the surviving ruins of earth and stone recall the magnificence and splendor of ancient Persia.

Borobudur

Java, Indonesia

Elevated on an earthen plateau overlooking forests and rice fields in the surrounding valley, the impressive Buddhist pilgrimage shrine of Borobudur rises against a dramatic background of smoking volcanoes. Resembling a low truncated pyramid in silhouette, the spectacular monument is a stone-clad earthen hill recalling mythical Mount Mehru. Eight platforms of diminishing sizes rise from a plinth measuring 500 feet (150 meters) square to a height of 116 feet (35 meters). In the Mahayana Buddhist cosmic system, the plinth symbolizes the underworld, the lower five platforms represent the world of man, and the upper three recall the world of God. Together the nine platforms represent the stages of life that lead to Nirvana. Walled galleries embellished with bas-reliefs line the perimeter

Hollow stupas containing outward-gazing statues of Buddha encircle the shrine's upper levels. Photo by Digital Vision Ltd./SuperStock.

PLATFORMS

A large stupa crowns Borobudur's nine platforms that symbolize the stages of life leading to Nirvana. Photo by Brian Brake.

of the lower platforms, while 72 hollow stupas, each containing an outward-gazing statue of Buddha, encircle the upper three terraces. A large stupa crowns the apex. Recalling the design of a mandala, the monument's biaxially symmetrical plan has steep stairways along the cardinal axes. Probably built between 778 and 842 A.D., the magnificent structure was abandoned and overgrown until its rediscovery in the nineteenth century. Recent investigations indicate that extensive earth fill was required to level the original site. Having survived periodic monsoons and occasional earthquakes for more than a millennium, the sophisticated stupa of Borobudur has proven to be one of the most enduring achievements of Indonesian art and architecture.

Preah Vihear

Cambodia

Terminating in a spectacular cliff-top view, the causeway of Preah Vihear extends for more than half a mile (800 meters) through the jungles of northern Cambodia. Courtesy of L'Ecole Française d'Extrême Orient.

Perched dramatically on the edge of a 2,000-foot (600-meter) high cliff in northern Cambodia, the Khmer temple complex of Preah Vihear consists of several shrines built on platforms along a processional causeway more than half a mile (800 meters) long. The extraordinary complex is situated in the rugged Dangrek Mountains that rise abruptly from a flat plain along Cambodia's northern border. Preah Vihear's causeway lies along a north-south axis terminating in spectacular views from the cliff's crest to the south. From the base of the entry stairs to the north, pilgrims ascend 150 feet (46 meters) to the platform of the entry shrine. Continuing along an inclined ramp that leads through an intermediate shrine, they finally arrive at the sacred monastery elevated some 480 feet (146 meters) above

PLATFORMS

the level of the entry. Although striking views await visitors at the cliff's edge, the monastery purposely looks inward for contemplation. The monastery's tower once rose more than 70 feet (21 meters) above the secluded courtyard. From the monastery portal, pilgrims may gaze northward across the mountainside that gently descends toward the nearby jungles of Thailand. Construction of the extraordinary ensemble began in the latter ninth century A.D. and continued for three centuries with additions by several kings. Meaning "sacred monastery" in the Khmer language, Preah Vihear is built on a sacred mountain that pilgrims ascended because the experience offered spiritual rewards and the solitude necessary for religious meditation. (N.B. Preah Vihear also is spelled Prah Vihear.)

Emerald Mound

Natchez, Mississippi

One of the largest and best-preserved ancient earthworks in eastern North America, Emerald Mound creates an impressive ceremonial platform some 33 feet (10 meters) above surrounding grade. Photo by William N. Morgan.

Extending some 770 feet (235 meters) along a natural ridge in southeastern Mississippi, the impressive earth platform known as Emerald Mound is the second largest earthwork in pre-Columbian North America. Measuring at most 770 feet (235 meters) in length at its base, well-preserved Emerald is exceeded in volume of earth fill only by Monks Mound at Cahokia some 500 miles (800 kilometers) to the north (see page 158). Emerald Mound's platform rises about 33 feet (10 meters) above the surrounding terrain and seems to have been flanked originally by three small truncated mounds along both its north and south sides with a larger seventh truncate at its east end. The summit of the restored west pyramid towers perhaps 65 feet (20 meters) above the adjacent grade, affording visitors panoramic views over the encircling treetops. The site lies near the south end of the Natchez Trace, not far from the pre-Columbian town of Anna that was abandoned about the time Emerald began to prosper. The enormous structure incorporates the remains of a former village that had rectangular wood buildings; village debris and earth fill form the steep sides of the present earthwork. Burials with exotic artifacts that were found in the east mound suggest burial ceremonialism possibly related to the six house mounds flanking the ceremonial plaza. Forerunners of the Natchez Indians apparently used the site between 1300 and 1600 A.D., with major construction occurring after 1500.

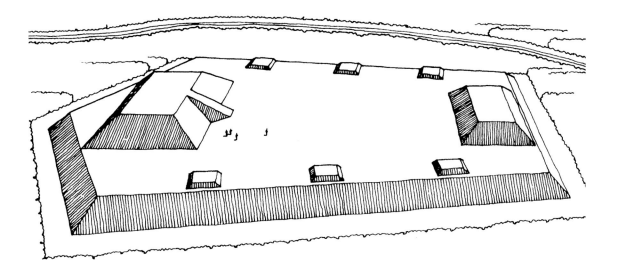

Native Americans carefully placed eight truncated mounds on top of the grand platform, affording views of the Mississippi River 6 miles (10 kilometers) to the west above the treetops. Drawing by William N. Morgan.

Governor's Icehouse

Williamsburg, Virginia

The observation platform surmounting the icehouse terminates a formal garden walk that begins in the service court of the Governor's Palace. Courtesy of Colonial Williamsburg.

PLATFORMS

Taking maximum advantage of the earth's inherent insulating characteristic, the icehouse of the Governor's Palace in Williamsburg is encased in superimposed pyramids of earth. Anchoring the northwest corner of the palace's ten-acre (four-hectare) garden, the impressive earth structure is surmounted by an observation platform from which visitors may gain commanding views to the south of the palace, its outbuildings, formal gardens, and terraces descending to the banks of Capitol Landing Creek, a tributary of the nearby York River. On the upper truncated pyramid's north side an inconspicuous brick-arched portal leads via a short passageway to the barrel-shaped ice storage vault lined with brick. Measuring more or less 9 feet (2.7 meters) in diameter and descending some 15 feet (4.6 meters) into the earth, the vault is capped by a brick dome and rests on a sump of generous proportions. The structure was constructed in the early 1700s with minor repairs in the 1930s, clearly demonstrating the enduring qualities of brick and earth as building materials. From the icehouse a formal axis leads south through a maze of holly shrubbery, a garden with fruit-bearing trees, a hedge-enclosed graveyard with a weeping willow at its center, and a box garden divided into quadrants and shaded by live oaks. A photograph taken during the restoration process in the winter of 1935 clearly depicts the truncated pyramids of earth in a comprehensive view from the southeast.

Insulated by earth and lined with brick, the centuries-old Governor's Icehouse is encased within truncated pyramids of earth. Courtesy of Colonial Williamsburg.

Dallas–Fort Worth Airport

Grapevine, Texas

One of the largest earth-moving projects in history has been the construction of the Dallas–Fort Worth Airport—involving 25 million cubic yards (19 million cubic meters) of earthwork in its initial phase alone. Its completed runways will be, in effect, level platforms measuring 20,000 by 300 feet (6,100 by 90 meters) and capable of bearing the weight of aircraft that are several times heavier than those presently in service. Located on a site larger than Manhattan between Dallas and Fort Worth, the airport formerly was gently rolling agricultural and pastoral land surrounded by gullies and dry washes that drain the terrain in all directions. Situated well away from houses, schools, and hospitals, the busy airport has a capacity in the range of 200,000 passengers per day. The functional design is based on the concept of several small airports, each with a plane at its gate, parking at its front door, and convenient ticketing in between. Each terminal is semicircular in plan with terraced parking located centrally; aircraft are assigned to the perimeter for maximum maneuvering space. The master

Ramps and bridges extend over expressways while vehicles and trains interconnect parking facilities, terminals, and cities. Photo by R. L. Mitchell.

plan consists of pairs of terminals located on opposite sides of a road system that serves arriving passengers on lower levels and departing passengers on upper levels. Three bridges over the expressway permit planes to taxi from one side of the airport to the other, while automated trains interconnect the terminals below grade. Construction was expedited by precasting concrete components in nearby factories and assembling them on site.

Constructed by superhuman machines on a scale to match, the Dallas–Fort Worth Airport converts pastures with gullies and washes into level platforms. Photo by R.L. Mitchell.

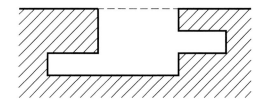

Excavations

Troglodyte Settlement, 4200 B.C., Safadi, Negev, Israel

Berber Village, Matmata, Tunisia

Zhog Tou Community, Luoyang, Honan, China

Guadix Community, 1500 to present, Granada, Spain

East Bank Bookstore, University of Minnesota, 1973,
    Minneapolis, Minnesota

Beach Cabañas, 1980s, Baja California Sur, Mexico

The immense open-pit mines of Chuquicamata in northern Chile have produced copper ore since pre-Columbian times. Courtesy of the Anaconda Company.

For the purpose of this discussion, excavations refer to the systematic removal of material from the earth. One of the world's largest excavations is the immense open-pit mine at Chuquicamata in northern Chile from which copper ore has been extracted since pre-Columbian times. Systematic mining operations began in 1915; five decades later the gaping mine exceeded 2 miles (3 kilometers) in length and half a mile (.8 kilometers) in width, an awesome demonstration of the capability of contemporary industrial technology to shape the earth.

EXCAVATIONS

Several hundred miles north of Chuquicamata in the rugged Andes Mountains of Peru, Inca builders and their predecessors long ago created five unusual excavations, each consisting of as many as fourteen terraces. Known as Moray Gardens, the site is located near Ollantaytambo. The terraces typically are about 6 feet (2 meters) high and vary greatly; they apparently were used for agricultural experimentation, taking advantage of different microclimates at various levels. If it had been used for ceremonialism or as an amphitheater, the largest bowl might have accommodated as many as 60,000 spectators.

Far from South America, another type of agricultural experimentation has been developed in man-made excavations resembling funnels in the sand dunes of northeastern Algeria. Here, in the Souf oases, productive date palm trees are grown in small groves that are excavated deeply enough into the dunes to reach the constant water table. The craters are protected from encroaching sand dunes by woven palm fences and dry

The Incas used the irrigated earthworks of Moray Gardens for agricultural experimentation on various levels. Courtesy of Werner-Gren Foundation for Anthropological Research.

stone walls. The undulating excavations of the Souf oases extend some 25 miles (40 kilometers) across the North African sand dunes.

Solving a difficult site problem with an inventive solution, the undergraduate library of the University of Illinois was detached from the main library and placed underground near the middle of the campus in the 1960s. Excavated two stories below the surrounding grade and oriented toward a central courtyard, the building preserves the open character of the main north-south mall of the campus and avoids casting shadows on a field nearby that is used for agricultural research.

More than six thousand years ago, early settlers in the Negev Desert of Israel excavated chambers and tunnels into the desert floor, perhaps as protection against marauders or for relief from extreme temperatures and sandstorms. Apparently for similar reasons, the Matmatas of Tunisia long ago learned to excavate courtyards into the desert floor and carve subterranean chambers on two levels around the walls to serve as living, sleeping, activity, and storage rooms. Sloping entry tunnels interconnect the courtyards with the desert floor.

Palm fences and stone walls protect sand craters with date palms from encroaching dunes in the Souf oases of Algeria. Photo by Georg Gerster.

EXCAVATIONS

Reading rooms and stacks overlook a central courtyard in the undergraduate library at the University of Illinois in Urbana, Illinois. Courtesy of Richardson, Severens, and Scheeler Associates, architects.

In addition to underground housing in Israel and Tunisia, communities of similar dwellings accommodate large numbers of people who live today in China and Spain. Subterranean residences in Zhog Tou, for example, are easily carved into the loess soil of northern China where residents enjoy all the comforts of home, privacy, and moderate temperatures in both summer and winter. Similar advantages attract people who live underground in the arid environment of Guadix in southern Spain; here electronic gadgets and television aerials seem to be at least as numerous as the residents.

Institutional and commercial buildings in recent years also have found advantages in creating space below grade. The University of Minnesota's new bookstore and office building introduces daylight through skylights and a triangular courtyard deep into the interior spaces below a busy pedestrian concourse traversing the campus. Far to the south on the warm Pacific coast of Mexico lies an attractive beachfront resort; here in Cabo San Lucas an inconspicuous group of cabañas is recessed into the coastal dunes with uninterrupted views of the sea.

Troglodyte Settlement

Safadi, Negev, Israel

Safadi is one of several early troglodyte settlements in the southern Negev Desert that is dug into banks of a dry riverbed called a wadi. Courtesy of Centre de Recherche Française de Jérusalem.

Early settlers in the semiarid Negev Desert chose to live underground, carving domical chambers, tunnels, and entryways into the hard hillside soils overlooking streams and drainage ways. The underground chambers may have served as a defense against marauders or possibly as protection from the heat of the day, the chill of the night, and the biting windborne sands of desert storms. Arriving as early as 6,500 years ago, the settlers successfully grew barley, wheat, and lentils and raised cattle, sheep, and goats in an area where rainfall was more plentiful and the climate milder than it is today. At Safadi, tunnels descended as deeply as 23 feet

EXCAVATIONS

(7 meters) below the surface and interconnected as many as ten chambers in a single group. At the nearby troglodyte settlement of Shiqmim, hundreds of underground chambers have been found ranging in size up to 13 by 26 feet (4 by 8 meters) in plan with ceilings as high as 8 feet (2.4 meters). That the residents processed grains, produced pottery vessels, carved bowls from imported basalt, stored foods in underground silos, manufactured metal tools and shell jewelry, and carved ivory figurines from elephant tusks indicates extensive trade activity. With the development of a flood plain beginning around 4200 B.C., the people began to build rooms above grade, using the subterranean chambers increasingly as storerooms, particularly for grains.

Dating back more than 5,000 years, Safadi is a complex of linked caves that are connected by tunnels, ventilated by occasional airshafts, and entered through roofless rooms. Courtesy of Centre de Recherche Française de Jérusalem.

Berber Village

Matmata, Tunisia

Almost invisible in the barren landscape of central Tunisia, the Berber community of Matmata is built below grade for protection against the extreme heat and cold of the desert climate. For thousands of years the inhabitants have lived in grottoes consisting of central courtyards with subterranean chambers carved into the surrounding walls. Constructing

In Matmata, stone-faced storage lofts and living spaces open into the central courtyard where children play, meals are prepared, and rainwater is collected. Photo by Kidder-Smith.

EXCAVATIONS

a typical residence begins by excavating a shaft perhaps 33 feet (10 meters) square to a depth of about 26 feet (8 meters) into the desert floor. Excavated material deposited around the edges of openings accounts for the uneven village terrain. A gently sloping entry tunnel then is dug from the ground level down to the courtyard. Serving as an entry court, light well, storage area, kitchen, and play yard for the surrounding chambers, the central space also contains a cistern where occasional rainwater is stored. Rooms on two levels traditionally are carved into the soft sandstone or clay walls around the court. Lower-level chambers serve as living, sleeping, or activity rooms, while the compartments above serve as storage lofts accessible by stairs or ladders. Well-insulated and whitewashed interior rooms often have furniture alcoves or shelving recessed into their walls. More elaborate façades sometimes are faced with stones or bricks with arched portals. Guest quarters and livestock shelters are located along the entry tunnel. Before tourism developed, Matmatans traditionally eked out meager livings by producing barley, figs, olives, and date palms.

Protected against the extreme heat and cold of the Tunisian desert, the subterranean dwellings are barely visible in the barren landscape. Photo by Kidder-Smith.

Zhog Tou Community

Luoyang, Honan, China

A brick-lined passageway with gentle steps descends into a typical underground dwelling in Zhog Tou Community. Photo by Newton Morgan.

Rectangular courtyards serve as outdoor living areas, gardens, or utility yards, providing daylight and ventilation. Photo by Newton Morgan.

Easily carved into the wind-borne loess soil of northern China, the subterranean dwellings of Zhog Tou accommodate perhaps half of the community's 5,000 residents. A typical underground dwelling is entered by means of a 6-foot (2-meter) wide, brick-lined passageway with gentle steps that lead down about 30 feet (9 meters) to the south, turn to the west, and continue down to a landing some 30 feet (9 meters) below grade. Visitors then pass through a short tunnel and enter into the corner of an atrium measuring 38 feet (12 meters) square and open to the sky. Five arched portals lead from the atrium into rooms carved out of surrounding walls; these are arranged so that several more chambers could be added in the future. Serving variously as living, sleeping, storage, or food-preparation

An arched portal and short tunnel expand the dwelling size to accommodate domestic requirements. Photo by Newton Morgan.

Carved in the soft loess soil, barrel-vaulted ceilings and whitewashed walls are typical of the subterranean chambers. Photo by Newton Morgan.

spaces, the rooms are barrel vaulted with ceilings rising at most 9.5 feet (3 meters) above the floor. The chambers are roughly twice as long as they are wide and range in area from 130 to 175 square feet (12 to 16 square meters). Plaster composed of silt, straw, and mud underlies the whitewashed interiors. Wood doors and frames resting on raised stone sills provide the only openings into the well-insulated rooms. The kitchen stove is the only source of heat in winter; summer temperatures usually hover around 70 degrees Fahrenheit (21 degrees Celsius). Brick wainscots and tile caps protect the atrium's walls from erosion, and low brick walls at the tops of the atrium and entryway surround the openings on the village level above.

Guadix Community

Granada, Spain

The largest concentration of troglodytes in Europe, the Guadix community has thousands of dwellings with walled entry courts leading to chambers excavated into the sandstone hillside. Photo by Newton Morgan.

Excavated into the sandstone hillsides of an arid landscape, the dwellings of Guadix are home to the largest concentration of troglodytes in Europe. Situated about 12 miles (20 kilometers) north of Spain's highest mountain range, the Sierra Nevada, Guadix lies some 35 miles (56 kilometers) east of the provincial capital of Granada. Here in the roughly one-square-mile (2.6-square-kilometer) district known as Barrio Santiago, more than 2,000 underground households accommodate almost half of the town's 30,000 residents. Living below grade offers the advantage of year-round temperatures in the range of 66 degrees Fahrenheit (19 degrees Celsius) in a zone of extreme winter and summer surface temperatures. Typical subterranean chambers are rectangular in plan with widths of 10

Typically ventilated by airshafts, windowless chambers often have vaulted, whitewashed ceilings and light-colored walls for maximum light reflectivity. Photo by Newton Morgan.

to 13 feet (3 to 4 meters) and often have vaulted ceilings up to 10 feet (3 meters) high. Basic services include electricity and water; many dwellings have television aerials, and some have connections for telephone and electronic communications. Short tunnels link chambers together to provide the desired number of living rooms, bedrooms, and support spaces. The interiors of more elaborately finished residences often include marble or carpeted floors, wainscots and extensive millwork. Providing ventilation for the windowless underground chambers, airshafts that resemble whitewashed chimneys project above the ground. Entry courts of richly diverse designs distinguish individual dwellings.

East Bank Bookstore

Minneapolis, Minnesota

Continuous skylights introduce daylight deep into the multi-story interior spaces of the East Bank Bookstore, admitting the winter sun and deflecting the summer sun. Courtesy of Meyers & Bennet, architects/BRW.

Built almost entirely below the level of the surrounding site, the East Bank Bookstore / Records and Admissions Building preserves the views of two historic buildings and accommodates a busy pedestrian artery in the heart of the University of Minnesota campus. A broad concourse traverses the 192- by 204-foot (59- by 62-meter) site diagonally; on one side the glass wall of a continuous skylight introduces daylight deep into the multistory interior spaces below. On the other side of the walkway, a triangular garden descends through several stories of offices, with concrete window boxes arranged to admit the winter sun and deflect the summer sun. Sloping glass walls introduce day lighting to the lowest floor 42 feet (13 meters) below the campus level. Light-filled two- and three-story-high

Flanked by a skylight on one side and garden views on the other, a broad concourse diagonally traverses the roof of the new bookstore and offices amid traditional neighbors. Courtesy of Meyers & Bennet, architects/BRW.

interior volumes enrich the spaces of the bookstore and offices. Consuming less than half the energy of a similar above grade building, the thermal efficiency of the facility is being monitored closely to assist in the design of future underground structures. Surface temperatures in Minneapolis range from more than 90 degrees Fahrenheit (32 degrees Celsius) in summer to well below freezing in winter, while temperatures 10 feet (3 meters) below ground level hover around 50 degrees Fahrenheit (10 degrees Celsius) throughout the year. Constructed of exposed concrete columns with board-formed walls and deeply coffered concrete slabs, the building was completed in 1977 at a cost slightly lower than a conventional structure would have cost.

Beach Cabañas

Baja California Sur, Mexico

Recessed into the beachfront dunes along the southern tip of Baja California Sur, an inconspicuous group of cabañas enjoys spectacular views of the seashore without destroying its natural environment. Located at the junction of the desert landscape and the sea, Cape San Lucas lies at the confluence of the Gulf of California with the Pacific Ocean. The cape affords limited protection for a small harbor accommodating fishing boats, pleasure craft, and, occasionally long ago, English pirates lying in wait for Spanish galleons. Barely visible from afar, the beach cabañas are situated on two levels, with the inshore units looking over the sand-covered rooftops of those closer to the beach. Stairs lead down through sunlit gardens to the lower level. Built of concrete and covered with earth, each cabaña

Excavated into the beachfront dunes of Cabo San Lucas, an inconspicuous group of cabañas enjoys unobstructed views of the seashore without destroying the environment. Courtesy of Ricardo Legorreta Arquitectos.

EXCAVATIONS

Barely visible along the shore, the cabañas are situated on two levels; each has a private entryway and secluded patio overlooking the Pacific Ocean. Courtesy of Ricardo Legorreta Arquitectos.

has a sliding glass wall overlooking the sea. Retaining walls extend toward the beach, forming private sitting and sunbathing patios for individual units. When the exterior doors are closed, not even the sound of the surf can be heard due to the acoustical properties of the earth and masonry materials enclosing the cabañas. In addition to the advantages of conserving the natural environment and offering a high degree of acoustical and visual privacy from nearby neighbors, the beach cabañas also are relatively economical to construct and maintain. The structures are highly energy efficient as well due to the insulating characteristic of the earth that largely precludes the need for air conditioning associated with conventional structures built above grade.

 # Modified Earth

Casa Grande, Caliche, 1300–1450, Coolidge, Arizona

Glaumbaer Farms, 1880s, Glaumbaer, Iceland

Great Mosque, Mud Brick, 1906–1907, Djenné, Mali

Visions of Paolo Soleri, 1970s, Scottsdale, Arizona

Earth-dome Housing, 1980s, Hesperia, California

Studio and Courtyard, Rammed Earth, 1990s, Tucson, Arizona

Durable and inexpensive utility buildings near Oruro, Bolivia, traditionally are constructed with turf blocks reinforced by roots. Photo by Frederico Ahlfeld.

MODIFIED EARTH refers to adaptations of earth to new architectural uses. These include drying mud bricks in the sun, combining soil with straw or grass, compacting earth under pressure, adding water to earth to achieve plasticity, or combining different soils for increased strength. Illustrating these principles, blocks of soil containing undisturbed grass roots can be removed from the earth and stacked in walls before drying out to create rudimentary structures. Undisturbed roots improve the soil's durability and tensile characteristics. Sod block structures have been developed successfully in areas of grassland with low rainfall like Iceland, the Great Plains of North America, and the Altiplano of South America, where the inhabitants in Bolivia traditionally build and use sod block shelters in daily life.

Early settlers on the Great Plains of North America built sod block houses of increasing sophistication during the latter part of the nineteenth century. Typical of these was the house that Sylvester Rawding built for his family during the 1880s in Custer County, Nebraska. Complete with double-hung windows and hinged wood doors, the durable sod house was warm in the winter, cool in the summer, well protected against wind storms and marauders, unaffected by prairie fires, rot resistant, economical to build, and impervious to insects. During the early twentieth century wood-frame houses began to replace sod houses on the Great Plains, not because they were more livable or more practical, but as a status symbol demonstrating the owner's improved economic condition.

MODIFIED EARTH

Aiming to create architecture for the poor, Egyptian architect Hassan Fathy advocated traditional methods of constructing buildings with mud brick arches, vaults, domes, screens, and walls. Of the new towns that he designed, the best known probably was New Gourna, a compact community intended for 7,000 residents on 50 acres of land near Luxor. Due to their adobe construction, the buildings were cool in the summer, warm in the winter, and exceptionally economical to construct. Fathy designed the village square, market, mosque, and individual houses (including his own) according to the needs and preferences of the residents.

The sod house built in 1886 by the Sylvester Rawding family demonstrates a successful method of coping with environmental adversities. Courtesy of Nebraska State Historical Society, S. D. Butcher Collection.

Thick adobe walls reinforced by massive buttresses enclose the mission church of San Francisco de Assisi near Taos, New Mexico, a favorite subject of architects, photographers, and artists, including Georgia O'Keeffe. Built in the late 1700s by members of the Franciscan order, the church and its modest courtyard are surrounded by a low adobe wall. Based on a traditional Native American construction system, adobe consists of sun-dried bricks reinforced with straw and laid in courses to create walls or applied in layers over matted fibers on wood beams to build roofs. Adobe walls often are finished with adobe plaster that is periodically refurbished.

Modular mud-brick domes and arches flank the village square of New Gourna, one of several new towns in Egypt designed by the distinguished architect Hassan Fathy. Courtesy of the Aga Khan Trust for Culture.

Another material of earth architecture is caliche, a natural element underlying the surface of the desert floor. Used extensively in the construction of Paquimé, in northern Mexico (page 160), caliche also was the principal building material for Casa Grande, Arizona, where the walls increased from 4.5 feet (1.4 meters) in thickness at the base to 1.8 feet (.53 meters) at the parapet. Another type of modified earth architecture is found in the sod houses of Nebraska, Oruro, and Glaumbaer Farms in Iceland, where steep-roofed, interconnecting houses are built of turf blocks in an area where customary building materials are scarce or non-existent.

Built of sun-dried mud brick and finished by a coat of mud plaster, the Great Mosque of Djenné, Mali, recalls similar construction techniques used elsewhere in dry climates of Africa, the Middle East, and the Americas. In dramatic contrast with more conventional approaches to earth architecture, the visions of Paoli Soleri encompass entirely new cities and transportation systems that challenge existing social, political, and economic norms.

In recent years an economical modified earth-housing system has been developed in California based on filling fabric sleeves with earth to create continuous layers of sandbags. Relatively unskilled workers place one layer of sandbags on top of another in a circular plan that terminates

　　MODIFIED EARTH

in a dome. Barbed wire is laid between the sandbags to control shear. Another promising direction for earth architecture is rammed earth, a proven technology applied with imagination and conviction by architect Rick Joy for an office with courtyard in Tucson and in several other projects in Arizona.

Massive adobe buttresses reinforce the stout masonry walls of the mission church of San Francisco de Assisi, near Taos, New Mexico. Photo by Newton Morgan.

Casa Grande

Coolidge, Arizona

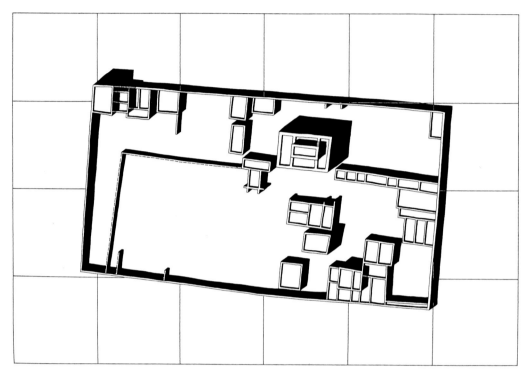

Probably built in the early 1300s near the center of a large, walled compound, Casa Grande overshadowed its more modestly scaled neighbors. Drawing by William N. Morgan.

Erected some 700 years ago in the desert area of south central Arizona, the impressive great house of Casa Grande was made of caliche, a subsoil of high lime content that occurs in natural deposits just below the surface of the desert. The three-story-high walls of the great house tapered from 54 inches (137 centimeters) in thickness at the base to 21 inches (53 centimeters) at the parapet. Layers of caliche mixed with water formed horizontal courses about 26 inches (65 centimeters) thick in the walls. Floor construction consisted of pockets recessed in the massive walls that received wood beams supporting a reed subfloor capped by a layer of caliche. Each floor contained five interconnecting rooms, all about 9 feet (2.7 meters) wide and ranging from 24 to 35 feet (7.3 to 10.7 meters) in length. Ladders provided vertical access to the rooms. A small penthouse centered on the rooftop afforded clear views of the mostly one-story-high buildings in the walled complex surrounding the great house. Known as the Hohokam, the desert farmers also constructed forty or so other complexes together with about 600 miles (960 kilometers) of irrigation canals in the vicinity of present-day Phoenix. The canals typically measured approximately 3 feet (.9 meters) in depth and 2 to 6 feet (.6 to 1.8 meters) in width. The Hohokam first appeared early in the Christian era and seem to have flourished until shortly before Spanish explorers entered the Southwest.

Hohokam builders erected the massive walls of Casa Grande using caliche, a naturally occurring subsoil rich in lime that is found just below the surface of the desert. Photo by Newton Morgan.

Glaumbaer Farms

Glaumbaer, Iceland

Turf mounds undulating above a plain of grass are defining features of Glaumbaer Farm along the north coast of Iceland near Akureyri. Faced with an acute shortage of large trees needed to construct conventional buildings, Icelanders created complex structures by interconnecting clusters of small, contiguous houses built of turf blocks. A typical Icelandic turf block building is Glaumbaer Farms, an assemblage of thirteen adjacent turf structures that were constructed over a period of several decades. Icelandic turf grows thickly, providing a strong, durable texture of roots and soil. Glaumbaer's turf structures are lined with imported wood and are insulated by thick earth walls and roofs. In an area of moderate rainfall, turf structures can last for a century, provided the roof slope is not too flat to shed water or too steep for turf to grow without drying and cracking during dry weather. Glaumbaer Farm's guest rooms, storage spaces, toolshed, and toilet open directly to the exterior. An exterior door also opens into a long hallway that interconnects the living quarters. These include a kitchen able to serve up to twenty residents, a large pantry, a cool room for meat and dairy products, a service entry with animal shelter, and a long living room where the farmer, his family, and hired hands ate, worked, and slept. Chambers accommodated the children at one end of the living room and the parents at the other. Windows provided light for women to card and spin wool. In more recent times reinforced concrete structures are replacing traditional turf buildings in Iceland.

Traditional turf buildings in Iceland often consist of contiguous rows of rhythmical, vaulted earthworks. Photo by Bernd Buier.

End walls sometimes are faced with imported wood planks with conventional windows and doors. Photo by Bernd Buier.

Faced with an acute shortage of building materials, Icelandic builders until recently utilized blocks of native turf, which grows thickly and provides durable root mats with favorable insulation characteristics. Photo by Sigríður Sigurðardóttir.

Great Mosque

Djenné, Mali

Superbly formed and highly expressive, the Great Mosque in Djenné is one of the world's most environmentally sustainable structures. Photo by age fotostock/SuperStock.

The largest mud-brick building in the world, the Great Mosque of Djenné, traces the origins of its architectural design to the eleventh century. The magnificent building is constructed on a rectangular plinth of sun-dried bricks that raises the structure above the floodplain of the nearby Bani River. Walls of mud brick, mortar, and plaster 16 to 24 inches (40 to 60 centimeters) thick enclose the mosque and subtly taper as they rise to support the distinctive upper walls, towers, and spires. The massive walls insulate the interior spaces during the heat of the day and cool down at night. Cooling is assisted by removing caps from roof vents at night to exhaust warm air from the interior. Although wood beams projecting from the façades may seem to be decorative features, they structurally support scaffolding for the replastering of exterior walls each spring and also minimize stresses caused by variations in humidity and temperature. Senior masons coordinate the annual replastering using mud mixed with rice husks that form a material called *banco*. Rebuilt in 1906 and 1907, the Great Mosque dominates the large market square of Djenné. Founded about 800 A.D., the city is situated near the head of routes leading to gold and salt mines. A regional trading center in the sixteenth century, Djenné also became known as a center of Islamic learning and pilgrimages. In 1988 the old town of Djenné and its Great Mosque were designated a UNESCO World Heritage Site.

Wood beams projecting from the massive mud-brick walls of the Great Mosque serve to support scaffolding for annual replastering. Photo by SuperStock, Inc./SuperStock.

Visions of Paolo Soleri

Scottsdale, Arizona

Carefully reshaping the natural environment to meet the needs of human-kind, the visionary architect Paolo Soleri has proposed a number of com-pact, high-density cities called *arcologies*, a term combining the words "architecture" and "ecology." An example of the visionary cities is Arco-forte, an arcology that consists of fourteen multistory residential blocks with rooftop promenades grouped around an open civic center. Situated

Arcoforte, shown here in plan, is one of a series of visionary cities conceived by architect Paolo Soleri to meet the needs of humankind. Courtesy of Cosanti Foundation.

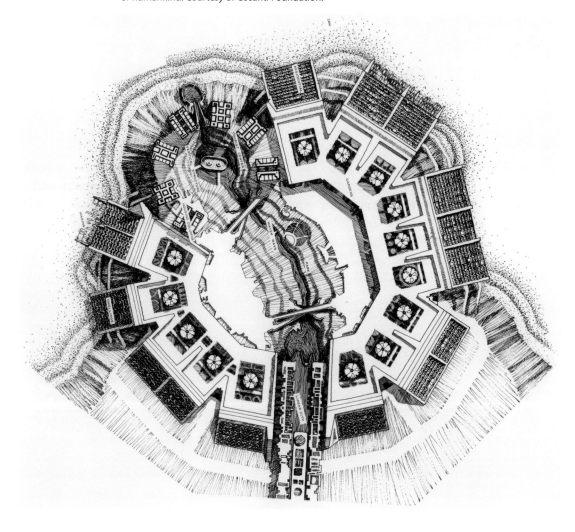

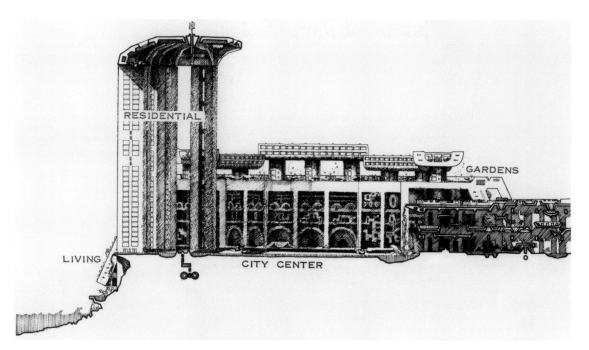

A seventy-story-high residential tower overlooks
the urban center of Arcoforte, a compact pedestrian
city that eliminates sprawl and reduces pollution.
Courtesy of Cosanti Foundation.

in a canyon along a rugged coastline, Arcoforte has, below its civic center, grottoes filled with automated industries related to fishing. Seven slender 70-story-high housing towers support a domical roof above the tourist center. Educational and tourist facilities reflect Soleri's view that travel, education, and the exchange of ideas are integral components of urban life. Elevators, escalators, and moving sidewalks preclude the need for streets or automobiles, and high-speed transit systems interconnect Arcoforte with neighboring cities. Urban sprawl and pollution are eliminated; nature surrounds the city. In another of the architect's arcologies, Arcoindian I, an apartment tower surrounds an atrium above several levels of industry. Intercity transit tubes leave the tower's base above playgrounds along the lake. Arcoindian II, Theodiga, and Theology arcologies explore variations in sites with differing topographies, climates, and populations. Although Soleri has completed several more modest projects involving earth architecture, none exceeds the promise of his visionary cities.

Earth-dome Housing

Hesperia, California

Extensive research into traditional earth-building systems led Iranian architect Nader Khalili to develop Eco-Dome, an exceptionally economical method of earth construction derived from the site itself. The basic technique involves filling a fabric sleeve with earth to create a continuous layer of sandbags placed one above another in a circular plan. A single strand of barbed wire is laid between each layer to prohibit shifting and to improve earthquake resistance. Successive layers corbel inward as they rise, forming a dome at the top of the structure. Arches can be created in

Sandbags filled with earth are laid in a plan of contiguous circles. Courtesy of Khalili/ Cal-Earth Eco-Dome.

Successive courses are corbelled inward near the top to form an enclosing dome. Courtesy of Khalili/ Cal-Earth Eco-Dome.

MODIFIED EARTH

Exterior surfaces are filled in and
smoothed to receive waterproofing.
Courtesy of Khalili/Cal-Earth Eco-Dome.

the curving shell wall to provide small windows, niches, portals, or vaults
connecting with adjacent shells. Limited quantities of cement or lime may
be added to the earth fill to enhance the structural characteristics of the
components. Finish coats of stucco render the corrugated coil surfaces
smooth and watertight. Cool in the summer and warm in the winter, a
typical earth-dome house built in Hesperia, California, contains approximately 400 square feet of interior space, including a central living room
with four apses accommodating a kitchen, bathroom, bunkroom, and
foyer with storage. Windows are oriented to admit maximum sunlight in
winter, and a towerlike wind scoop is situated for optimum ventilation.
Particularly suited to arid environments, earth-dome housing units are
fireproof, nontoxic, and environmentally sound; they require no timber
or steel and no special skills for construction. New communities of earth-
dome housing have been proposed for California and Iran.

Studio and Courtyard

Tucson, Arizona

The richly textured rammed-earth walls of Rick Joy's studio consist of three kinds of earth chosen to emphasize the colored patterns of the horizontal layers. Courtesy of Rick Joy, architect.

Windowless rammed-earth walls surround architect Rick Joy's recently built studio incorporating a long, narrow courtyard in a historic district of Tucson. The introspective studio is illuminated by a glass wall that overlooks the courtyard to the north and by a linear skylight along the conference room's east wall. Following a traditional building technique for semiarid climates, the massive rammed-earth walls range in thickness from 18 to 48 inches (46 to 122 centimeters) and contain sufficient thermal capacity to cope with the temperature extremes of the desert Southwest. Spread footings and concrete stem walls elevate the earth walls slightly above grade. The exposed, richly textured walls consist of three kinds of earth chosen to emphasize the colored patterns of horizontal layers; small

The massive rammed-earth walls are sufficiently thick
to cope with the temperature extremes of the desert
Southwest. Courtesy of Rick Joy, architect.

amounts of iron oxide further enhance the natural beauty of the walls.
Three percent Portland cement is added for additional strength, and limited structural steel components are utilized as required. Layers of earth 10
inches (25 centimeters) thick are placed between standardized formwork
normally used in concrete construction. Each layer is rammed down to a
thickness of 5 inches (12.5 centimeters), achieving 95 percent compaction
before the succeeding layer is placed. Finished floors are polished concrete slabs with thickened edges, while ceilings consist of reflective metal
panels with recessed incandescent lighting. Thick, rough-sawn wood entry doors and light maple furnishings reinforce the design's emphasis on
natural materials.

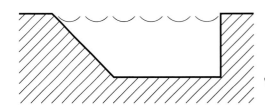 # Water Retained

Imperial Tomb, 300–710 A.D., Sakai, Osaka, Japan

Angkor Wat, Khmer, 1112–1152, Siem Reap, Cambodia

Katsura Gardens, 1615–1662, Kyoto, Japan

Khajou Bridge, Urban Promenade, 1657, Esfahān, Iran

Land Reclamation Polders, 1932 to Present, North Sea Coast, Netherlands

Nacimiento Dam, 1957, Paso Robles, California

I N THIS STUDY the category of water retained refers to bodies of water integrated with earthworks that are created to reshape the human environment. The bodies include moats, canals, reservoirs, lakes, harbors, dams, dikes, and land reclamation. For instance, double moats and six outward-projecting bastions surround the star-shaped city of Naarden located southeast of Amsterdam. One of several fortified cities once forming a line of defense around Amsterdam, earthworks, palisades, ramparts, and ravelins strengthened Naarden in three phases between 1350 and 1685. The surrounding moats were connected to a network of canals in order to

Multiple moats and projecting earth bastions fortify the historic city of Naarden in the Netherlands. Photo by KLM Aerocarto.

WATER RETAINED

Undulating earth walls retain an immense water reservoir for
the pumped storage power plant in Vianden, Luxembourg.
Courtesy of Société Générale pour l'Industrie Geneva.

resupply the city by sea or land during a siege. Present-day Naarden reflects the influence of Renaissance town planning with a cathedral marking its urban center.

Undulating walls of earth retain the Vianden reservoir in Luxembourg.
The reservoir supplies water to one of the world's largest pumped storage
power plants. Completed in 1964, the earth walls rise more than 50 feet (15
meters) above the surrounding ground level and are surfaced with rocks
and a layer of concrete coated with bitumen for watertight integrity.

Developers in several Florida cities have dredged huge areas of low-
lying terrain, including former mangrove swamps, in the process of creating luxury housing plots. Boat owners find the waterfront plots to be particularly desirable. During the 1950s peninsulas were developed on both
sides of the bay in the center of Naples along the Gulf of Mexico. Access to
individual plots on the peninsula requires extended streets that terminate
in cul-de-sacs. Over the years similar developments have been undertaken
on both sides of the Intracoastal Waterway in Fort Lauderdale, situated on
Florida's Atlantic coast. Recent environmental regulations have curtailed
substantially the development of wetlands in Florida.

Environmental regulations now preclude converting
wetlands into housing sites like these in Naples, Florida.
Courtesy of Naples Chamber of Commerce.

On the opposite side of the world, Japanese builders long ago excavated
moats around often-immense keyhole-shaped earthworks that contain
the tombs of Emperor Nintoku and other ancient rulers of Japan. Now
surrounded appropriately by buffers of trees, the revered burial monuments are set apart from their urban environments in Sakai. A different
sense of reverence comes across at the ancient Cambodian capital of Angkor, now shrouded in an encroaching jungle. Here during the eighth century Khmer builders constructed an extensive network of dikes, canals,
reservoirs, and earthworks that included special islands for such pilgrimage shrines as Angkor Wat.

Near Kyoto during the seventeenth century, the earth was reshaped
and a river was diverted to form the extraordinary Katsura Villa with its
contemplative gardens, one of the greatest masterpieces of Japanese architecture. Here the relationship between human beings and the natural
world seems complete; heaven and earth are very nearly one.

WATER RETAINED

Celebrating the city of Esfahān (Isfahan), the remarkable Khajou Bridge serves not only as a river crossing but also as a promenade for evening strollers, a marketplace for merchants, a place for women to dye wool, and occasionally as a dam to create a glimmering lake for festivals. Built of sun-dried bricks on stone piers, the remarkable two-level bridge is one of the foremost examples of Persian art, architecture, and urban planning.

Twentieth-century earth dams illustrate the potential of present-day technology for reshaping the earth on exceptionally large scales. A series of dams serving as dikes along the North Sea coast have facilitated the formation of polders, thereby increasing significantly the land area of the Netherlands. For very different reasons, an earth dam across the Nacimiento River in central California has created an immense reservoir that provides opportunities for recreation, water for neighboring communities, irrigation for farmlands, flood control, and power generation serving a substantial number of consumers.

In the 1950s dredged fill and canals reshaped the waterfront along the Intracoastal Waterway in Fort Lauderdale. Photo by Carroll Seghers.

Imperial Tomb

Sakai, Osaka, Japan

In Sakai, three moats filled with water surround Emperor Nintoku's immense earth tumulus that dates to the fifth century. Photo by Shoiche Umehara.

Emperor Nintoku's tomb is the largest and one of the oldest of Japan's remarkable keyhole-shaped tumuli. Containing more than 49 million cubic feet of earth, the enormous earthwork measures some 1,600 feet (500 meters) in length and 1,000 feet (300 meters) in width. The circular portion of the structure holds the tomb; it rises to a height of 115 feet (35 meters) from a base 800 feet (245 meters) in diameter. Slightly lower in height, the trapezoidal portion may once have been the site of mortuary ceremonies. Surrounding the immense tumulus are a broad inner moat and two smaller outer moats, all filled with water. Earth excavated from the moats seems likely to have been a major source of fill material for the vast earthwork. The great size of Nintoku's tomb indicates not only a relatively high level of technological sophistication but also a well-organized and highly motivated labor force capable of completing the immense project

successfully. Burial mounds containing the tombs of people of high social rank are called *kofun* in Japan; the Kofun Period (circa 300–710 A.D.) was the time of Japan's greatest mound-building activity. Constructed during the fifth century, the tomb of Nintoku is located in a cluster of 92 large and small kofun within an area of roughly 6.2 square miles (16 square kilometers) in the city of Sakai. Today a dense buffer of trees encircle Nintoku's tomb, setting the site apart from the surrounding urban environment.

Similar in scale to the Great Pyramid of Khufu in Egypt, Nintoku's keyhole-shaped tomb lies serenely in a now bustling modern city. Photo by Shoiche Umehara.

Angkor Wat

Siem Reap, Cambodia

Earth dug from the broad surrounding moat elevates the island of Angkor Wat above flood level. Photo by age fotostock/SuperStock.

Founded on an extensive network of dikes and canals linked to immense reservoirs, the Khmer capitol of Angkor represents the culmination of ancient architecture and a high point of civilization in Southeast Asia. Evolving from building traditions of the earlier kingdoms of Funan and Chenla, the Khmers developed magnificent monuments such as Angkor Wat, which is encircled by a broad moat. Here a wall 3.5 miles (5.6 kilometers) long surrounds a rectangular island containing three successive terraces of earth in its center that form a pyramid recalling mythical Mount Mehru.

WATER RETAINED

The highest of the five towers crowning Angkor Wat rises 213 feet (65 meters) above the surrounding terrain. The architectural achievements of the Khmers were made possible by a strong centralized government capable of creating and maintaining highly sophisticated irrigation systems. Canals dug into the jungles converted swamps into agricultural plots yielding up to three crops annually, thus assuring economic stability for the empire. Waterways also provided a source of fish for the burgeoning population, and a means of transportation that interconnected Angkor Wat with the Mekong Delta and coastal ports of Southeast Asia. Shaping the earth to elevate causeways, islands, and building foundations above flood waters, Khmer architects created an ingenious building system uniquely appropriate to their environments. Constructed by perhaps as many as 50,000 workers, twelfth-century Angkor Wat remains today an impressive monument to its creators.

Khmer builders successfully converted their low-lying terrain into a network of canals, causeways, agricultural plots, and building platforms. Photo by Michele Burgess/SuperStock.

Katsura Gardens

Kyoto, Japan

Expressing the interrelation of man and nature, the gardens and structures of the Katsura Imperial Villa are among the best-known and greatest masterpieces of Japanese architecture. Here between 1620 and 1652 the earth was shaped to form three different islands in a large pond with irregular shorelines. The reconfigured terrain presents ever-changing views as visitors stroll through the gardens. A variety of bridges and paths interconnect the islands, where visitors find four small viewing pavilions, several teahouses, and belvederes at various locations within the elegantly beautiful gardens. Extensive plantings highlight the four seasons in a complex microcosm of nature. Carefully situated in harmony with the overall

Seemingly a timeless work of nature, the beautiful Katsura gardens actually consist of three artificial islands created in a carefully shaped pond. Photo by William N. Morgan.

WATER RETAINED

Constantly changing views of the ponds and
gardens present strollers with elegantly beautiful
landscape scenes. Photo by William N. Morgan.

plan, the rooms of the main house open onto porches that overlook the
gardens to the south. Located on a 17-acre (7-hectare) site along the west
bank of the Katsura River near Kyoto, the estate has served as a country
retreat for aristocrats since the early 1600s. In contrast to the design of the
Katsura gardens, those of such sixteenth- and seventeenth-century European gardens as the Villa d'Este and Versailles assert man's dominance
over nature, imposing a symmetrical order on land and plants along a
central axis. At Katsura, however, man is considered to be a part of uni-
versal nature, including such features as hills, rocks, trees, rivers, flowers,
and clouds. Created by reshaping and embellishing the earth centuries
ago, the Katsura gardens continue today to enhance our understanding
of the interaction of man and nature.

Khajou Bridge

Esfahān, Iran

An urban esplanade and dam as well as a river-crossing, the beautifully embellished Khajou Bridge is one of the most remarkable structures in the world. Capable of converting the Zayandeh River into an urban lake along the south side of Esfahān (Isfahan), the bridge's sluice gates control the water level. Earthen levees along the riverbanks contain a lake suitable for royal barges on special occasions. In 1657 Shah Abbas II built the bridge of sun-dried bricks, tile, and stones that form stout piers supporting rhythmical arches. Following a natural rock outcropping across the river, the bridge consists of twenty small islands in a row separated by narrow channels that create the pleasant sound of splashing water. Downstream to the east women often dye fabrics in the flowing stream, fishermen occasionally try their nets, and sheep graze along the sandbars. Designed on two levels with royal pavilions at midspan, the bridge accommodates on its upper level a bustling street with vendors' stalls flanked by outward-

When its sluice gates are closed, the Khajou Bridge dams the river and creates a lake near the center of Esfahān.
Photo by William N. Morgan

WATER RETAINED

The foundations of the bridge are twenty small islands following a natural stone ridge across the river. Photo by William N. Morgan.

viewing alcoves. Measuring 433 feet (152 meters) in overall length, the Khajou Bridge lies along the ancient caravan route from Shiraz far to the south to the nearby bazaar of Esfahān to the north. Snowmelt in the distant mountains to the southwest is the source of water for the Zayandeh River, although the water flows mainly in the winter and spring and flooding is rare. A network of canals distributes water throughout the city, creating a large, tree-shaded oasis in the arid environment of Esfahān.

Land Reclamation Polders

North Sea Coast, Netherlands

Schockland Island was dredged from a shallow lake in 1936, reclaiming land from the sea. Courtesy of Netherlands Information Service.

By 1940, reclaimed land surrounded the former island, creating a new polder. Courtesy of Netherlands Information Service.

Masters of land reclamation and planning, the Dutch in recent times have increased the area of their densely populated nation by more than 10 percent. The reclaimed land formerly lay below areas of the Zuider Zee and the low-lying delta along the Netherlands' southwest coast. Proceeding according to a plan envisioned late in the nineteenth century, a dam was completed across the former Zuider Zee in 1932. Built of earth and rubble, the dam facilitated the construction of several huge polders containing in all some 550,000 acres (220,000 hectares) of arable land and creating an immense freshwater reservoir, Lake Ijssel. Polders require the building of dikes around the areas to be reclaimed, such as the 56-mile (90-kilometer) long earth dam around East Flevoland. As fertile alluvial soil is dredged up from the bottom of the Zuider Zee, such former islands as Schockland are absorbed into newly created farmlands. Sand, clay, peat, and other dredged materials produce varied colors in the new landscape that is

An immense dam of earth and rubble fill separates a vast freshwater reservoir, Lake Ijssel, from the saline North Sea. Courtesy of Netherlands Information Service.

laced with navigable canals and irrigation ditches. Along the southwest coast, North Sea storms caused frequent flooding in the delta of the Maas, Waal, and Lower Rhine rivers until a network of dikes was completed in the 1970s. One of these, the Veersche Gat dam, was built of shale, sand, and asphalt covered with stones, concrete, and grass. Newly created Lake Delta supplies freshwater to fields, factories, and towns, while the dikes provide new parklands and improve the national highway system.

Nacimiento Dam

Paso Robles, California

Used for power generation, irrigation, water supply, recreation, and flood control, 215-foot (66-meter) high Nacimiento Dam is built of earth, gravel, and stones laid in compacted layers. The huge earth structure spans the Nacimiento River in the Santa Lucia Range about 12 miles (19 kilometers) northwest of Paso Robles. Rising from a base at most 1,125 feet (343 meters) wide, the dam measures 1,470 feet (448 meters) in length at its crest and contains more than 3.4 million cubic yards (2.6 million cubic meters) of fill materials. The claylike core of the dam resists the passage of water; it consists of a wall of impervious fill diminishing from almost 200 feet (60 meters) in width at its bedrock foundation to 20 feet (6 meters) at its crest. The core's diminishing width reflects the diminishing water pressure from the bottom of the reservoir to the top. Immense berms of pervious shell on both sides of the dam hold the core in place and protect it from erosion. At the north end of the dam an ungated concrete spillway relieves potential floodwaters before they reach the crest. In the valley of oaks and pines west of the dam, Lake Nacimiento rises to a height of 800 feet (244 meters) above sea level. Covering an area of 8 square miles (20.7 square kilometers), the lake develops 165 miles (264 kilometers) of shoreline with many coves and inlets and has a healthy population of game fish. Since the dam's completion in 1957, the lake has provided areas suitable for camping, fishing, swimming, boating, and other recreational activities.

WATER RETAINED

The Nacimiento Dam is built of earth, gravel, and stone laid in compacted layers over a massive claylike core that resists the passage of water. Courtesy of Engineering News-Record.

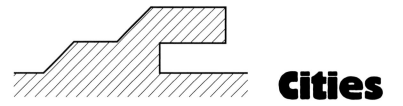 **Cities**

Poverty Point, 1730–1350 B.C., Floyd, Louisiana

Cahokia, Regional Center, 900–1250, Collinsville, Illinois

Paquimé, Trading Center, 1275–1500, Chihuahua, Mexico

Taos Pueblo, 1450, Rio Grande Valley, New Mexico

Urban Nucleus, 1965, Sunset Mountain Park, California

Dam Town Proposal, 1966, Pikeville, Kentucky

An enigmatic four-tiered mound rises above a plaza marking the center of the ancient city of Cuicuilco that flourished in central Mexico between 500 B.C. and 100 A.D. Photo by George and Audrey DeLange.

CHARACTERISTICS OF CITIES include relatively dense populations, urban planning, and often impressive institutional structures. In the Valley of Mexico a city apparently emerged about 500 B.C. at Cuicuilco on the southern outskirts of present-day Mexico City. Here a circular earth structure perhaps 370 feet (115 meters) in diameter rose some 75 feet (23 meters) above the surrounding plaza. Faced with random masonry, the Cuicuilco earthwork marks the center of an extensive urban zone that included residences, plazas, and avenues bordering small reservoirs, irrigation canals, fortifications, and related features. The circular structure seems to have been abandoned around 100 A.D. after lava from a nearby volcano buried much of the site. Today, intensifying urban sprawl threatens to destroy the remains of the ancient city.

Encircled by double earth walls that were separated by a 115-foot (35-meter) wide moat, the fortified city of Firuzabad was the first capital of the Sassanian empire (224–642 A.D.). Entrances at the cardinal points once passed through the 16–foot (5-meter) high perimeter walls that measured 4 miles (6.4 kilometers) in circumference. At Firuzabad's center a masonry tower almost 100 feet (30 meters) high bears traces of a circular stair that leads up to the flat roof where a fire once burned, perhaps as a signal to distant travelers or as a sacred link between the citizens and their Zoroastrian faith.

More than a millennium later in North America, Hopi builders created a new group of towns on the craggy mesa tops of northeastern Arizona. Following the Pueblo Revolt of 1680, many Hopi communities left their valley-floor sites and established highly defensible new towns in the nearby uplands. Built mostly of uncoursed stone and adobe materials taken from the site, the communities' cubelike room blocks typically were tightly grouped around compact central squares. Incorporating ma-

Encircled by a broad moat between earth walls in south central Iran, Firuzabad was the first capital of the Sassasian Empire. Photo by Georg Gerster.

Founded about 1700, Shipaulovi, like other Hopi towns, is built of indigenous stones and adobe materials taken from its mesa-top site. Photo by Newton Morgan.

jor rock outcroppings into rectilinear building masses, floor levels often varied from room to room, and kivas were chiseled into the sandstone depressions along the craggy edges of the mesas. Today the fiercely independent Hopi continue to live closer to their ancient traditions than do other Native Americans.

Addressing the problem of an increasing population in a limited land area, a team of Dutch architects in the 1960s proposed the Pampus Plan, an ambitious urban development for 350,000 residents on newly created islands to be dredged from the lake west of Amsterdam. A multilevel transportation artery for vehicles and a monorail would extend from the historic city through the new land so that no resident would be more than five minutes away from rapid transportation. The plan envisioned new schools, playgrounds, parks, boat basins, shops, offices, and churches surrounding apartment blocks up to twenty-four stories high. The spatially rich new community would offer diversity for Amsterdam's burgeoning population.

One of the oldest surviving earth cities in the Western Hemisphere, 3,700-year-old Poverty Point combines concentric rows of berms, truncates, and an immense effigy mound into a geometrically precise plan. Some 2,000 years later and far to the north of Poverty Point, Cahokia, the

largest pre-Columbian city in North America, included some 120 truncates, cones, ridge mounds, platforms, and other earthworks that were grouped around orderly plazas.

Two great cities of earth represent southwestern North America in this study: Paquimé is built of caliche, and Taos is built of adobe. The vast trading center of Paquimé contained a central marketplace, a ring of urban parks, numerous multistory buildings up to seven stories high, ball courts, public baths, an aqueduct, and a reservoir. Multistory Taos is one of the best-preserved Native American pueblos in the Rio Grande Valley. The traditional pueblo consists of multifamily room blocks with shaded exterior work areas arranged around a large plaza on both sides of a flowing creek.

Two recent proposals for cities in mountainous settings envision designs to reshape the earth with sensitivity while accommodating the needs of humankind. The Urban Nucleus proposes to reshape a mountaintop near Santa Monica to form a compact urban center that occupies only 3 percent of the land available. In the eastern United States, an earth dam with sculpted terraces for community functions typifies one of a series of promising new towns in the valleys of the Appalachian Mountains. Here numerous dams have been built to control floods and to generate electrical power, but none has explored the potential of the earth to create cities for human use and enjoyment.

The visionary Pampus Plan proposed an expanded city on newly created islands west of Amsterdam. Courtesy of Van den en Bakema.

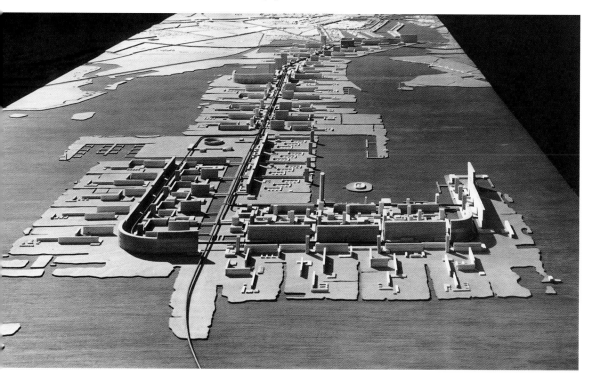

Poverty Point

Floyd, Louisiana

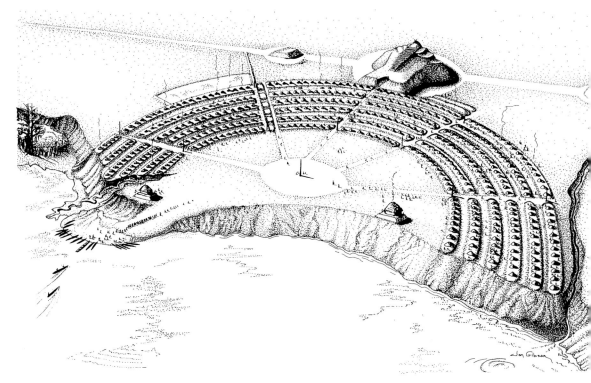

Created some 3,500 years ago, the vast and enduring
earthworks of Poverty Point continue to impress
visitors today. Drawing by Jon Gibson.

Built more than 3000 years ago by placing one basketful of earth upon
another, the immense ceremonial center known as Poverty Point extends
for about three-quarters of a mile (1.2 kilometers) along a high bluff in
northeastern Louisiana. Six rows of concentric curving earth platforms,
each about 5 feet (1.5 meters) high and 100 feet (30 meters) wide, enclose
a semicircular plaza oriented toward the east. The roughly 60-foot (18-
meter) wide spaces between the curving platforms provided fill material
for the earthworks. Radiating avenues divide the earth platforms into six
residential districts. Overlooking the complex to the west is an extraordinary effigy mound some 70 feet (21 meters) high that is shaped to resemble
a bird with a wingspread exceeding 640 feet (195 meters). Situated several
miles west of the Mississippi River, the ancient complex apparently was

constructed by Native Americans between 1730 and 1350 B.C. The tradition of creating earth architecture on a monumental scale seems to have begun in northern Louisiana as early as 5500 B.C. Partially destroyed by plowing during the nineteenth century, Poverty Point now is protected by law and has been designated as one of UNESCO's three world heritage sites in the United States. Once the center of a trading network that extended for hundreds of miles along the Mississippi Valley and beyond, Poverty Point today reminds us of the earth's enduring capacity to shape the environment of humankind.

A scaled plan of the ancient earthworks presents the once-grand ceremonial and trading center of Poverty Point at its zenith. The plan's grid is 660 feet (200 meters) square. Drawing by William N. Morgan.

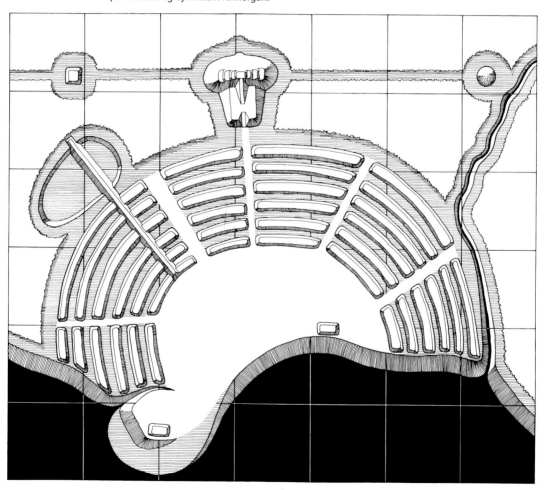

Cahokia

Collinsville, Illinois

Earth was the principal building material for the 120 truncated pyramids and geometric earthworks of Cahokia, the largest pre-Columbian city in North America. Located across the Mississippi River from present-day St. Louis, the city was the center of a trading network that extended from the Great Lakes to the Gulf of Mexico and from the Great Plains to the Appalachian Mountains. By 1200 A.D. the impressive city extended almost 3 miles (4.8 kilometers) from east to west and nearly 2 miles (3.2 kilometers) from north to south. Dominating the palisaded central area of Cahokia, multiterraced Monks Mound towered 100 feet (30 meters) above the grand plaza and was surmounted by a tall wood structure. Containing an estimated 22 million cubic feet (615,000 cubic meters) of earth, the immense truncate was the largest earth structure north of the pyramids at Teotihuacán and Cholula in central Mexico. An earth ramp extended from Monks Mound south to the plaza level where an athletic field and marketplace were flanked by ceremonial mounds and earthworks. Twin

The largest earthwork in pre-Columbian North America, multiterraced Monks Mound rose almost 100 feet (30 meters) above the grand plaza of Cahokia. Courtesy of Cahokia Mounds State Historical Site, Illinois.

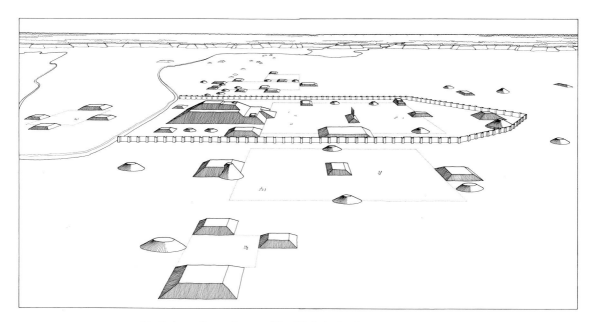

Located near the confluence of the Mississippi and Missouri rivers, Cahokia contained in all some 120 truncates, cones, ridge mounds, platforms, and other earth structures. Drawing by William N. Morgan.

mounds at the south end of the palisade enclosure consisted of a truncated charnel house mound to the east of a conical burial mound. By 1250 A.D. Cahokia's population may have numbered 15,000 residents, arts and crafts flourished, sociopolitical systems became highly developed, and architecture and planning attained high levels of sophistication. Thereafter Cahokia began to decline, though earth cities continued to appear elsewhere in eastern North America.

Paquimé

Chihuahua, Mexico

Built primarily of an earth material known as caliche, the once magnificent city of Paquimé was the pre-Columbian center of a trading network that extended from the Gulf of Mexico to the Pacific Ocean and from central Mexico to lower Colorado. Situated at an altitude of 4,860 feet (1,480 meters) above sea level, the urban complex encompassed an area of some 88 acres (36 hectares), of which only a relatively small portion has been excavated. Major components of the ancient city include ceremonial burial mounds, a large marketplace lined with merchants' booths, a ring of urban parks, sophisticated apartment buildings up to seven stories high, and a grand plaza dividing the city into halves. The vast complex incorporated such impressive elements as grand ramps and stairways, colonnaded public halls, ceremonial entryways, multiple ball courts, and spacious courtyards flanked by multistory buildings. An aqueduct conveyed water from the distant Sierra Madre Occidental to a reservoir near the urban center. From this point stone-lined channels lead under the

Magnificent Paquimé incorporated a large market square, a ring of urban parks, ball courts, a central plaza, an aqueduct, sweat baths, and apartments as high as seven stories. Photo by Newton Morgan.

public market and continue to the apartment blocks, sweat baths, and other structures to the south and east. Evolving from a modest hamlet of mud-domed houses built in pits during the ninth century, Paquimé apparently reached its period of greatest architectural efflorescence during the fifteenth century. The grand trading center subsequently fell into decline and probably was abandoned not long before early Spanish explorers began to find their way into the Southwest.

Abandoned shortly before Spanish explorers entered the Southwest, the caliche structures of Paquimé presently are being stabilized and partially restored. Photo by Newton Morgan.

Taos Pueblo

Rio Grande Valley, New Mexico

Sun-dried adobe brick walls with stucco finishes and wood-framed roofs formed the rectangular rooms of Taos and other Rio Grande Valley pueblos. Photo by Newton Morgan.

In many ways characteristic of traditional Native American architecture in the Rio Grande Valley, Taos Pueblo was constructed of sun-dried adobe brick walls with stucco finishes and flat, wood-framed roofs forming rectangular rooms. Native Americans traditionally entered the rooms by means of ladders and roof hatches that also served to emit smoke from interior hearths. Doors and windows in exterior walls are not characteristic of traditional architecture before the Spanish entered the Southwest. Early in the Christian era, pit houses began to emerge as a building type in the Southwest, perhaps because of the lack of construction materials other than a plentiful supply of earth and a limited supply of timber. Recessed into the earth several feet, pit houses are relatively easy to build

At Taos between 1000 and 1250 A.D., multistory room blocks replaced pit houses that were partially recessed into the earth. Photo by Newton Morgan.

and economical to maintain. Earth walls provide insulation against high summer temperatures and low winter temperatures, while the arid climate requires limited maintenance of exposed stucco surfaces. Shaded work areas and storage rooms above grade enhance livability. Taos grew from a pithouse settlement around 1000 A.D. into an early multifamily surface room block by A.D. 1250. By the 1500s Spanish explorers found Taos Pueblo, which then consisted of room blocks up to five stories high, arranged around a large plaza on both sides of a flowing creek. On festival days, rooftops served as bleachers for spectators watching athletic events, symbolic dances, and community-wide ceremonies.

Urban Nucleus

Sunset Mountain Park, California

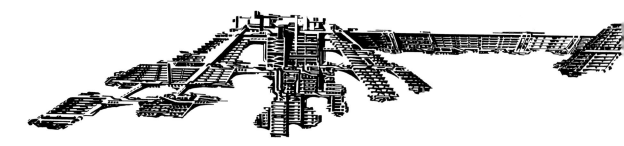

The Urban Nucleus proposes a compact new city that would reinforce the spectacular character of the mountainous terrain rather than destroy it. Courtesy of Cesar Pelli with A. J. Lumsden DMJM.

Perched on a promontory in Sunset Mountain Park near Santa Monica, the Urban Nucleus proposes a compact new town that would follow closely the site's rugged contours. Designed to occupy less than 3 percent of the mountainous terrain, the community would consist of low-rise structures that would reinforce the character of the land rather than destroy it. The Urban Nucleus would accommodate 7,200 dwelling units on platforms stepping down the hillsides from a town center that would be centrally located on the promontory. From a nearby highway residents would drive into a multilevel parking garage under the town center where they would leave their vehicles and proceed via elevators up to the urban plaza. Here they would find shops, restaurants, theaters, hotels, commercial and professional services, chapels, municipal offices, schools, childcare facilities, a heliport, and other urban services within convenient walking distances. Pneumatic supply tubes would convey items purchased in the town center or parcels dropped off in the garage directly to townhouses to await the arrival of residents. Inclined elevators and moving sidewalks would carry residents to their townhouses where they would have the best views of nature and little view of one another. No structure would cast a shadow on its neighbor or block its views. Projected costs for compact contour–rise construction are significantly lower than those for the conventional suburban sprawl that presently scars the landscape.

Designed to occupy only 3 percent of the mountainous site,
the pedestrian city radiates from its urban service center.
Courtesy of Cesar Pelli with A. J. Lumsden DMJM.

Dam Town Proposal

Pikeville, Kentucky

Since the 1930s dams and reservoirs have been constructed in the valleys of Appalachia to control floodwaters and, in some locations, to produce hydroelectric power. Carrying this precedent forward, a hypothetical community integrated with a new dam and lake was proposed in 1966 by John Ray, who at the time was a graduate student in architecture. The site of the proposed town was a dam then being built near Pikeville by the Army Corps of Engineers to control flooding on the lower Big Sandy

The proposed Pikeville dam would provide employment and urban amenities for a small city without destroying the environment. Courtesy of John Ray, University of Kentucky School of Architecture.

CITIES

Hydroelectric and flood-control dams offer opportunities to accommodate increasing populations in concert with magnificent and often rugged terrain. Courtesy of John Ray, University of Kentucky School of Architecture.

River. Ray envisioned a dam shaped to accommodate housing, parks, schools, stores, offices, and related facilities for a town with a population of perhaps 50,000 residents. The city hall and a large civic plaza would occupy the dam's summit, overlooking descending pools, waterfalls, and terraces that step down the sculpted valley walls below. Designed for incremental growth, the town eventually would become a city with possibly 100,000 residents, large enough to support a major hospital and cultural institutions, with industrial areas located downstream. In addition to the project near Pikeville, eight other sites in the region were identified for the potential development of dams with linear cities following the sinuous courses of valleys downstream. By terracing the dams and valleys, new towns could be created with the potential of enhancing flood control, developing sustainable sources of electric power, controlling suburban sprawl, and preserving the natural beauty of the mountainous terrain.

Afterword

Taken as a whole, the ninety examples of earth architecture cited in this study suggest ecologically sustainable relationships between human beings and their environments. The sites range geographically from Tahiti in the South Pacific to Iceland in the North Atlantic, and from a subterranean community in China to underground houses in Tunisia. Chronologically, the study extends through time from the troglodytes of the Negev Desert around 4200 B.C. to contemporary university libraries below grade, and to future earth cities envisioned for the mountains of North America and for the shores of the North Sea.

In terms of sustainability, earth architecture endures well over extended periods of time with little or no maintenance. For instance, about 3,700 years ago the precisely curving earth platforms of Poverty Point were built to a height of roughly 5 feet (1.5 meters) above grade to serve as house foundations. Today the earth platforms remain very much in evidence, but all vestiges of the wood houses originally built on them vanished long ago. A second example of sustainable earth architecture in northern Louisiana may be found at nearby Watson Break, where the earthworks are some 2,000 years older than those of Poverty Point, although the older site lacks the geometric sophistication and vast scale of the more recent one.

During the last four decades, photographs and data describing several hundred examples of earth architecture have been assembled and considered for inclusion in one of the study's nine categories. Reducing the possibilities to ninety of the best examples has been a daunting task, based primarily on geographical, chronological, and cultural diversity as well as the availability of relevant graphics and reliable data. For many of the sites included in this study, one or more equally worthy sites have been set aside to await presentation elsewhere in the future.

During the fifty years since this study began, several movements have developed in the field of architecture with little, if any, effect on earth architecture. During the 1960s, Postmodernism explored complexity and contradiction in architecture, essentially broadening the scope of Modernism without altering its course. In the late 1980s, Deconstructionism set aside accepted norms in favor of arbitrariness and novelty abetted by computers, leading to not surprising results. The turn of the twenty-first century ushered in a new movement concerned with such subjects as conservation of natural resources, renewable energy alternatives, passive

Native American builders created the ancient island city of Calos in southwestern Florida by depositing immense quantities of shell and refuse in the shallow waters of Estero Bay over a period of many centuries. Spanish visitors in the 1500s observed that a grand canal bisected the city and led to several smaller canals and boat basins. Flanking the grand canal were extraordinary mounds for the rulers of Calos, an elevated platform for the houses of the nobles, and numerous ridges and terraces where commoners lived, built boats, and mended nets.

solar potentials, and related topics that are remarkably similar to those that gave rise to the initiation of this study half a century ago.

The study's brief discussion of two proposed cities explores a number of principles of earth architecture in consonance with urban design. The Urban Nucleus looks outward across rugged terrain toward surrounding mountains with occasional glimpses of the horizon while the Dam Town looks inward to a valley of sculpted earth terraces and watercourses created by reshaping the natural terrain. Situated on a spectacular promontory, the Urban Nucleus envisions a compact new town in which residents enjoy a high degree of privacy and convenience together with a broad range of public services and amenities without the need of automobiles for transportation. The Dam Town proposes an attractive city oriented toward a central valley that is filled with waterfalls and lakes, and flanked by housing terraces with integrated urban services. The town's dam generates electricity and provides employment, while the central waterway provides convenient public transportation as well as a major amenity.

The designs of both communities rely on earth architecture directly or indirectly to conserve energy, control pollution, eliminate urban sprawl, minimize roads and utilities, reduce environmental impact, and encourage ecologically sustainable future expansion. Of particular relevance today, earth architecture is a promising field of study and a continuing source of inspiration for architects, planners, designers, ecologists, and others who are concerned with shaping the earth to create a sustainable balance between human beings and the world around them.

Appendix

Earth Samples

In the interest of comparatively analyzing soils from various locations in this study, thirteen samples were collected in the field and are recorded below. Generally speaking, a broad range of variables exists in terms of earth as a building material. Disregarding wind, rain, impurities, roots, and related conditions, the natural slopes of *cohesionless* materials usually range from 30 degrees for rounded particles, through sub-rounded and sub-angular, to 42 degrees for angular materials, based on a specific gravity of 2.7. *Cohesive* materials include those electrostatically or chemically bound (for example, calcium carbonate) and clays, which undisturbed maintain a natural slope in the range of 25 to 30 degrees around Boston, for example. Some materials change their characteristics after their original conditions are disturbed, such as the Cucaracha shale, which suddenly decreased its slope to 9 degrees some 60 years after the excavation of the Panama Canal.

In view of the broad variations inherent in soil characteristics, the advice of a competent soil mechanic is highly recommended for designing earth structures. Steve J. Poulos of the Harvard Soil Mechanics Laboratory provided guidance in the field of soil mechanics at the outset of this study; Dr. Poulos now is a private consultant in the Boston area.

Pachacamac

December 28, 1964. 18 miles south of Lima, Peru, on the Lurín River banks through dry coastal desert along the Pacific Ocean. 1.9" rainfall annually. Temperature range between 61° and 74° F., no vegetation.

Machu Picchu

December 30, 1964. Northwest of Cuzco on the Urubamba River in the Peruvian Andes, elevation 6,000'. 100" rainfall, temperatures range between 74° and 78° F. Tall tropical rain forest vegetation.

Chichén Itzá

January 3, 1965. Yucatán peninsula 75 miles east of Mérida, elevation 200'. 10" annual summer rainfall, semiarid climate, temperature average in summer 80° F. Tropical vegetation, semi-deciduous forest.

Uxmal

January 4, 1965. Yucatán peninsula 50 miles south of Mérida with conditions similar to Chichén Itzá.

Palenque

January 8, 1965. On the slopes of the Chiapas highlands, elevation 1,500', tropical rain forests. Climate, hot summer typically with temperatures averaging over 80° F.

Teotihuacán

January 11, 1965. On the central plateau near Mexico City, elevation about 7,000'. Average annual rainfall 26", semiarid with average annual temperature 58° F., natural vegetation mesquite, cacti, and yucca.

Acoma

April 18, 1966. 65 miles west of Albuquerque, New Mexico, altitude 5,000', annual rainfall 8", semiarid climate with temperatures ranging between 92° and 22° F., a tableland vegetated with mesquite and desert grass.

Baphuon

April 27, 1966. Siem Reap, Cambodia, in the Mekong River basin. Sea level tropical monsoon climate with annual rainfall average 58", tropical jungle.

Chandigarh

May 1, 1966. 100 miles north of Delhi, elevation 1,000', annual rainfall 20", warm climate with monsoons, natural vegetation, short grasses, and weeds.

Khajou Bridge

May 3, 1966. On the Zayandeh Rud River in Esfahān, Iran, annual rainfall 4.4", continental type climate with dryness and temperature extremes ranging between 30° and 100° F., vegetation arid grass and camel thorn.

Persepolis

May 4, 1966. Northeast of Shiraz, little rainfall, elevation 2,000', continental type climate with dryness and temperature extremes ranging between 30° and 100° F., no vegetation.

Petra

May 5, 1966. South of Amman, Jordan, on the Arabian Plateau, elevation 5,000', desert climate, hot days, cool nights, less than 10" annual rainfall, no vegetation.

Deir el Bahri

May 6, 1966. Luxor, Egypt, desert climate with temperatures ranging from cool to 107° F., almost rainless, no vegetation.

Bibliography

Alkire, William. *An Introduction to the Peoples and Culture of Micronesia.* 2nd ed. Menlo Park, California: Cummings, 1977.

Arthus-Bertrand, Yann. *Earth from Above.* New York: H. N. Abrams, 2002.

Aveni, Anthony F. *Between the Lines: The Mystery of the Giant Ground Drawings of Ancient Nasca, Peru.* Austin: University of Texas Press, 2000.

Baert, Brigitte, "Architectures Souterraines." *Decomag,* February 1984: 49–56.

———. "Une Architecture Naturale." *La Maison,* June 1984: 8–14.

Baggs, Sydney A. "Underground Architecture." *Australia House and Garden,* December 1977: 185.

Beardsley, John. *Earthworks and Beyond: Contemporary Art in the Landscape.* 3rd ed. New York: Abbeville Press, 1998.

Bellwood, Peter S. *Man's Conquest of the Pacific.* New York: Oxford University Press, 1979.

Betsky, Aaron. *Landscapers, Building with the Land.* New York: Thames & Hudson, 2002.

Birkerts, Gunnar. *Subterranean Urban Systems.* Ann Arbor: Industrial Development Division, Institute of Science and Technology, University of Michigan, 1974.

Boudron, David. *Designing the Earth.* New York: H. N. Abrams, 1995.

Brain, Jeffrey P. *Mississippian Settlement Patterns.* Studies in Archaeology Series. New York: Academic Press, 1978.

Brookes, John. *Gardens of Paradise.* New York: Meredith Press, 1987.

Brown, Percy. *Indian Architecture: Buddhist and Hindu Periods.* Bombay: D. B. Taraporevala and Sons, 1965.

Burger, Edmund. *Geomorphic Architecture.* New York: Van Nostrand Reinhold, 1986.

Carlson, John B. "America's Ancient Skywatchers." *National Geographic* 177.3 (1990): 76–107.

Codex Nuttal [also called Codex Zouche]. *Codex Nuttal: Facsimile of an Ancient Mexican Codex Belonging to Lord Zouche of Harynworth, England.* Intro. Zelia Nuttal. Cambridge, Mass., 1902.

Coe, Michael D. *The Maya.* 6th ed. New York: Thames & Hudson, 1999.

Cordell, Linda S., and George J. Gumerman, eds. *Dynamics of Southwestern Prehistory.* Washington, D.C.: Smithsonian Institution Press, 1989.

Craib, John L. "Micronesian Prehistory: An Archaeological Overview." *Science* 219 (1983): 922–27.

Davidson, Janet M. *The Prehistory of New Zealand.* Auckland: Longman Paul, 1984.

Davidson, Marshall B. *Lost Worlds.* New York: American Heritage, 1962.

Dethier, Jean. *Down to Earth.* London: Thames & Hudson, 1982.

Drexler, Arthur. *Transformations in Modern Architecture.* New York: The Museum of Modern Art, 1979.

———. *Twentieth Century Engineering.* New York: The Museum of Modern Art, 1964.

Earth Sheltered Housing Design. Minneapolis, Underground Space Center, 1978.

Edelhart, Mike. *The Handbook of Earth Shelter Design.* New York: Doubleday, 1982.

Edwards, I.E.S. *The Pyramids of Egypt.* London: Penguin, 1975.

Emory, Kenneth P. *Stone Remains in the Society Islands.* Bernice P. Bishop Museum Bulletin 116. Honolulu: Bernice P. Bishop Museum, 1933.

Fagan, Brian. *Ancient North America.* 2nd ed. New York: Thames & Hudson, 1995.

Fairhurst, Charles, ed. *Underground Space.* Minneapolis: University of Minnesota, May/June 1976.

Fathy, Hassan. *Architecture for the Poor.* Chicago: University of Chicago Press, 1969.

———. *Natural Energy and Vernacular Architecture.* Chicago: University of Chicago Press, 1986.

Folsom, Franklin, and Mary Etting F. Folsom. *America's Ancient Treasures*. 4th ed. Albuquerque: University of New Mexico Press, 1993.

Forde-Johnston, J., Jr. *History from the Earth*. London: Phaidon Press, 1974.

Frampton, Kenneth. *Studies in Tectonic Culture*. Cambridge, Mass.: MIT Press, 1996.

Gaprindashvili, Ghivi. *Vardzia*. Leningrad: Aurora Art Publishers, 1975.

George Nelson on Design. New York: Watson Guptill, 1979.

Giedion, Sigfried. *Space Time and Architecture*. 5th ed. Cambridge, Mass.: Harvard University Press, 1967.

Groslier, Bernard Phillips, and Jacques Arthaud. *Angkor: Art and Civilization*. London: Thames & Hudson, 1957.

Guidoni, Enrico. *Primitive Architecture*. New York: H. N. Abrams, 1978.

Gumerman, George J., and Emil W. Haury. "Prehistory: Hohokam." In *Southwest*. Vol. 9 of *Handbook of North American Indians*. Washington, D.C.: Smithsonian Institution, 1979. 75–90.

Gumerman, George J., David Snyder, and W. Bruce Masse. "An Archaeological Reconnaissance in the Palau Archipelago, Western Caroline Islands, Micronesia." *Southern Illinois Center for Archaeological Investigation Research Papers*, no. 23 (1981).

Gutkin, E. A. *Our World from the Air*. New York: Doubleday, 1952.

Hawkes, Jaoquetta. *Atlas of Ancient Archaeology*. New York: McGraw-Hill, 1974.

Heyden, Doris, and Paul Gendrop. *Precolumbian Architecture of Mesoamerica*. Trans. Judith Stanton. New York: H. N. Abrams, 1973.

Hudson, Charles L. *The Southeastern Indians*. Knoxville: University of Tennessee Press, 1978.

Jackson, J. B. *Landscapes*. Amherst: University of Massachusetts Press, 1970.

Jellicoe, Goeffrey, and Susan Jellicoe. *The Landscape of Man*. New York: Viking Press, 1975.

Jennings, Jesse D. *Prehistory of North America*. 2nd ed. New York: McGraw-Hill, 1974.

———. *Ancient North Americans*. New York: W. H. Freeman, 1983.

———, ed. *The Prehistory of Polynesia*. Cambridge, Mass.: Harvard University Press, 1979.

Kassler, Elizabeth B. *Modern Gardens and the Landscape*. New York: The Museum of Modern Art, 1964.

Keefe, Laurence. *Earth Building*. New York: Taylor and Francis, 2005.

Kennedy, Roger G. *Hidden Cities: The Discovery and Loss of Ancient North American Civilization*. New York: Free Press, 1994.

Kirch, Patrick Vinton. *On the Road of the Winds*. Berkeley: University of California Press, 2000.

Knowles, Ralph L. *Energy and Form*. Cambridge, Mass.: MIT Press, 1974.

La Nier, Royce. *Geotecture*. South Bend, Ind.: Royce La Nier, 1970.

Lumbreras, Luis G. *The Peoples and Cultures of Ancient Peru*. Washington, D.C.: Smithsonian Institution Press, 1979.

Martindale, David. *Earth Shelters*. New York: E. P. Dutton, 1981.

———. "New Homes Revive the Art of Living Underground." *Smithsonian* 9.11 (February 1979).

McCarter, Robert. *William Morgan: The Master Architect Series VII*. Melbourne: Images Publishing Group, 2002.

McHarg, Ian L. *Design with Nature*. Philadelphia: Falcon Press, 1971.

McQuade, Walter. "An Overlooked Bargain: Bright Young Men with Designs on the Future." *Fortune,* July 1966: 124–25.

Milanich, Jerald T., and Charles H. Fairbanks. *Florida Archaeology*. New York: Academic Press, 1980.

Millon, René. *Extensión y población de la ciudad de Teotihuacán en sus diferentes periodos: Un cálculo provisional*. Mexico, D.F.: Instituto Nacional de Antropología e Historia, 1976.

———. *El problema de la integración en la sociedad Teotihuacana*. Mexico, D.F.: Instituto Nacional de Antropología e Historia, 1967.

———. "Teotihuacán." *Scientific American* 216 (June 1967): 38–48.

———. "Teotihuacan: Completion of Map of Giant Ancient City in the Valley of Mexico." *Science,* 1970: 1077–82.

———. *Teotihuacán Map*. Austin: University of Texas Press, 1971.

———. *El valle de Teotihuacán y su contorno*. Mexico, D.F.: Instituto Nacional de Antropología e Historia, 1972.

Moreland, Frank L. *The Uses of Earth Covered Buildings*. National Science Foundation. Washington, D.C.: Government Printing Office, 1976.

Morgan, William N. *Ancient Architecture of the Southwest*. Austin: University of Texas Press, 1994.

———. *Precolumbian Architecture in Eastern North America*. Gainesville: University Press of Florida, 1999.

———. *Prehistoric Architecture in the Eastern United States*. Cambridge, Mass.: MIT Press, 1980.

———. *Prehistoric Architecture in Micronesia*. Austin: University of Texas Press, 1988.

Morris, James. *Butabu, Adobe Architecture of West Africa*. New York: Princeton Architectural Press, 2004.

Morris, Theodore. *Florida's Lost Tribes*. Gainesville: University Press of Florida, 2004.

Morrish, William Rees. *Civilizing Terrains*. San Francisco: William Stout Publishers, 1996.

Morrison, Samuel Eliot. *The European Discovery of America: The Southern Voyages, 1492–1616*. New York: Oxford University Press, 1974.

Nabokov, Peter, and Robert Easton. *Native American Architecture*. New York: Oxford University Press, 1989.

Ortiz, Alfonso, vol. ed. *Southwest*. Vol. 9 of *Handbook of North American Indians*. William C. Sturtevant, gen. ed. Washington, D.C.: Smithsonian Institution, 1979.

Panofsky, Erwin. *Meaning in the Visual Arts*. Garden City, N.Y.: Doubleday Anchor, 1957.

Park, Edwards, and Jean Paul Carlhian. *A View from the Castle*. Washington, D.C.: Smithsonian Institution, 1987.

Pfeiffer, John. "America's First City." *Horizon* 16.2 (1974): 58–63.

Phillips, Philip, James A. Ford, and James B. Griffin. *Archaeological Survey of the Lower Mississippi Valley*. Papers of the Peabody Museum of American Archaeology and Ethnology, Harvard University, 1951.

Prescott, William H. *History of the Conquest of Peru (1847)*. Ed. Victor W. von Hagen. New York: Mentor Books, New American Library of World Literature, 1961.

Robertson, D. S. *A Handbook of Greek and Roman Architecture*. London: Cambridge University Press, 1954.

Robertson, Donald. *Pre-Columbian Architecture*. New York: George Braziller, 1963.

Rowan, Jan C. "The Earth." *Progressive Architecture,* April 1967.

Rudofsky, Bernard. *Architecture Without Architects*. New York: The Museum of Modern Art, 1964.

Rudolph, Paul M. "Comments on Vernacular Architecture." In *Paul Rudolph, 1983–1984 Recipient of the Plym Distinguished Professorship*. Ed. James P. Warfield. Urbana-Champaign: School of Architecture, University of Illinois, 1983.

Rykwert, Joseph. *The First Moderns: The Architects of the Eighteenth Century*. Cambridge, Mass.: MIT Press, 1983.

Scott, Geoffrey. *The Architecture of Humanism*. Garden City, N.Y.: Doubleday, 1956.

Scully, Vincent. *Pueblo: Mountain, Village, Dance.* 2nd ed. Chicago: University of Chicago Press, 1989.

Smith, C. Ray. "The Earth." *Progressive Architecture*, April 1967.

Spreiregen, Paul D. *The Architecture of William Morgan*. Austin: University of Texas Press, 1987.

Squier, Ephraim G., and Edwin H. Davis. *Ancient Monuments of the Mississippi Valley*. Smithsonian Contributions no. 1, Washington, D.C., 1848.

Steele, James. *Ecological Architecture: A Critical History*. London: Thames & Hudson, 2005.

Sterling, Ray. *Earth Sheltered Housing Design I.* Minneapolis: University of Minnesota, 1978.

Sterling, Raymond, and John Carmody. *Earth Sheltered Housing Design.* 2nd ed. New York: Van Nostrand Reinhold, 1985.

Stevens, John L. *Incidents of Travel in Yucatan*. Vols. 1–2. New York: Dover, 1963.

Stuart, George E., ed. *Peoples and Places of the Past*. Washington, D.C.: National Geographic Society, 1983.

Stuart, George E., and Gene S. Stuart. *Discovering Man's Past in the Americas.* Washington, D.C.: National Geographic Society, 1969.

Sverbeyef, Elizabeth. "Computer Architecture House in the Earth." *House and Garden*, April 1976: 122–25. Rpt. in *The Earth Sheltered Handbook* (Milwaukee: Tech/Data Publications, 1980).

Teresi, Dick. *Lost Discoveries: The Ancient Roots of Modern Science, from the Babylonians to the Mayas.* New York: Simon & Schuster, 2002.

Thomas, Cyrus. *Report on the Mound Explorations of the Bureau of Ethnology.* Accompanying paper in the 12th Annual Report of the Bureau of Ethnology to the Secretary of the Smithsonian Institution, 1890–91, I-742. Washington, D.C., Government Printing Office, 1894.

Valadés, Adrian Garcia. *The City and Its Monuments.* Mexico, D.F.: Instituto Nacional de Antropología e Historia, 1976.

Waring, Anthony J., Jr., and Preston Holder. "A Prehistoric Ceremonial Complex in the Southeastern United States." *American Anthropologist* 47.1 (1945). Rpt. in "The Waring Papers," ed. Stephen Williams (Cambridge, Mass.: Harvard University, 1968).

Williams, Stephen. "The Aboriginal Location of the Kadohadacho and Related Tribes." In *Explorations in Cultural Anthropology.* Ed. Ward Goodnough. New York: McGraw-Hill, 1964.

———. "The Eastern United States." In *Early Indian Farmers and Villages and Communities.* Ed. W. G. Haag. Washington, D.C.: National Park Service, Department of the Interior, 1963.

———. "Settlement Patterns in the Lower Mississippi Valley." In *Settlement Patterns in the New World.* Ed. G. R. Wiley. Viking Fund Publications in Anthropology, no. 23 (1956).

———, ed. *The Waring Papers: The Collected Works of Antonio J. Waring, Jr.* Papers of the Peabody Museum, Harvard University, 1968.

Williams, Stephen, and Jeffrey P. Brain. *Excavations at Lake George, Yazoo County, Mississippi, 1958–1960.* Papers of the Peabody Museum, Harvard University, 1970.

Wood, Michael. *The World Atlas of Archaeology.* New York: Portland Press, 1985.

Wu, Nelson I. *Chinese and Indian Architecture.* New York: George Braziller, 1963.

Index

William N. Morgan has practiced architecture since 1960 in Jacksonville, Florida, where he has established an international reputation for design excellence. A Fellow of the National Endowment for the Arts and of the American Institute of Architects, he is an Eminent Scholar of the University of South Florida and has taught at the Harvard Graduate School of Design and other universities. He is the author of *Prehistoric Architecture in the Eastern United States* (1980), *Prehistoric Architecture in Micronesia* (1988), *Ancient Architecture of the Southwest* (1994), and *Precolumbian Architecture in Eastern North America* (1999).